OXFORD ENGLISH MONOGRAPHS

General Editors

THE SOCIAL
MISSION OF
ENGLISH CRITICISM

1848–1932

CHRIS BALDICK

CLARENDON PRESS · OXFORD

1987

Oxford University Press, Walton Street, Oxford OX2 6DP
Oxford New York Toronto
Delhi Bombay Calcutta Madras Karachi
Petaling Jaya Singapore Hong Kong Tokyo
Nairobi Dar es Salaam Cape Town
Melbourne Auckland
and associated companies in
Beirut Berlin Ibadan Nicosia

Oxford is a trade mark of Oxford University Press

Published in the United States
by Oxford University Press, New York

First published 1983
First issued as a paperback with corrections and addition 1987

British Library Cataloguing in Publication Data
Baldick, Chris
The social mission of English criticism
1848–1932.—(Oxford English monographs)
1. English literature—19th century—History
and criticism   2. English literature—
20th century—History and criticism
3. Literature and society
I. Title
820.9   PR451
ISBN 0–19–812821–5
ISBN 0–19–812979–3 (Pbk)

Library of Congress Cataloging in Publication Data
Baldick, Chris.
The social mission of English criticism, 1848–1932
(Oxford English monographs)
Bibliography: p.
Includes index.
1. Criticism—Great Britain—History. 2. English Literature—
History and criticism. 3. English literature—Study and teaching—
History. 4. Literature and society—Great Britain.
5. Social problems in literature.
I. Title.
PR63.B35 1983   801'.95'0941   83–4101
ISBN 0–19–812821–5
ISBN 0–19–812979–3 (Pbk)

Printed in Great Britain
at the University Printing House, Oxford
by David Stanford
Printer to the University

ACKNOWLEDGEMENTS

This book is based upon a doctoral thesis researched under the supervision and kind encouragement of Valentine Cunningham. I have benefited from comments upon previous drafts by Terry Eagleton, Peter Faulkner, and Patrick Parrinder, and from conversations with Terence Hawkes, John Simons, and my former students at the University of Exeter. From the shortcomings of this work all the above are, of course, absolved.

Permission to reproduce copyright material is acknowledged from the University of Michigan Press for extracts from R. H. Super's edition of *The Complete Prose Works of Matthew Arnold*; from the Literary Estate of Q. D. Leavis and from Chatto and Windus Ltd. for extracts from Q. D. Leavis's *Fiction and the Reading Public*; from the Literary Estate of Q. D. Leavis for extracts from F. R. Leavis's *For Continuity*; from the Cambridge University Press for extracts from George Sampson's *English for the English* and Arthur Quiller-Couch's *On the Art of Writing*; from Methuen and Co. Ltd. for extracts from T. S. Eliot's *The Sacred Wood*; from Routledge and Kegan Paul Ltd. for extracts from I. A. Richards's *Practical Criticism* and *Principles of Literary Criticism*. Extracts from T. S. Eliot's *Selected Essays* are reprinted by permission of Faber and Faber Ltd. In the United States, excerpts from *Selected Essays* by T. S. Eliot, copyright 1932, 1936, 1950 by Harcourt Brace Jovanovich; renewed 1960, 1964 by T. S. Eliot, 1978 by Esme Valerie Eliot: reprinted by permission of the publisher.

I would also like to thank Mary Laws, who patiently trained me in the use of the word processor on which I typed and corrected the drafts.

C. B.

CONTENTS

Abbreviations   ix

1. INTRODUCTORY: CRITICISM AND ITS HISTORY   1

2. MATTHEW ARNOLD'S INNOCENT LANGUAGE   18
   i   The future of poetry and the function of criticism   18
   ii Intellectual deliverance   26
   iii The missing centre   42
   iv Hellenizing consistently   49

3. A CIVILIZING SUBJECT   59
   i   The clown in the boudoir   63
   ii An additional accomplishment   67
   iii The native culture   70
   iv Confessions of a pimp   75

4. LITERARY-CRITICAL CONSEQUENCES OF THE WAR   86
   i   The alien yoke   87
   ii The Newbolt Report   92
   iii Immaterial communism   98

5. 'ON THE SIDE OF THE ARTIST': T. S. ELIOT'S EARLY
   CRITICISM 1917–1924   109
   i   Criticism and creation   113
   ii Simultaneous order   119
   iii Matter for argument   124

6. LITERARY-CRITICAL CONSEQUENCES OF THE PEACE:
   I. A. RICHARDS'S MENTAL LEAGUE OF NATIONS   134
   i   Chaos and its inhabitants   136
   ii 'The most insidious perversion'   142
   iii The most valuable people   147
   iv 'A rear-guard religious action'   156

CONTENTS

7. THE LEAVISES: ARMED AGAINST THE HERD   161
   i An anti-Marxist consensus   163
  ii Sociology of the herd   175
 iii The power of advertisements   186

8. A COMMON PURSUIT: SOME CONCLUSIONS   196
    i Science   199
   ii Theory   203
  iii Literacy and public controversy   205
  iv Whiggery and history   209
   v The problem of order: mental wholes   212
  vi The problem of order: the social organism   218
 vii Substitution and ideology   223
viii The practicality of criticism   239

Select Bibliography   237

Index   247

ABBREVIATIONS

Where abbreviation allows, page references for certain works frequently cited appear within the text in square brackets. The abbreviations for these works appear in the key below. Longer references and other notes are indicated numerically in the usual way, and appear at the end of each chapter.

CE   F. R. Leavis and Denys Thompson, *Culture and Environment: The Training of Critical Awareness* (1933).

CPW   Matthew Arnold, *The Complete Prose Works of Matthew Arnold,* ed. R. H. Super (11 vols., Ann Arbor, Michigan, 1960–77) (see Bibliography for titles of individual volumes).

FC   F. R. Leavis, *For Continuity* (Cambridge, 1933).

FRP   Q. D. Leavis, *Fiction and the Reading Public* (1932; Harmondsworth, 1979).

PC   I. A. Richards, *Practical Criticism: A Study of Literary Judgement* (1929).

PLC   I. A. Richards, *Principles of Literary Criticism* (1924; 3rd edn. 1928; reset 1967).

SAP   I. A. Richards, *Science and Poetry* (1926).

SE   T. S. Eliot, *Selected Essays* (1932; 3rd edn. 1951).

SW   T. S. Eliot, *The Sacred Wood: Essays on Poetry and Criticism* (1920; 7th edn. 1950).

1. INTRODUCTORY: CRITICISM AND ITS HISTORY

> The precept given by a wise man, as well as a great
> critic, for the construction of poems, is equally true
> as to states.
>   Burke, *Reflections on the Revolution in France*

In recent years much ink has been shed to the effect that a
'crisis' is besetting the study of English literature, particularly
in higher education. As the following chapters may help to
show, this is nothing new: from the very beginning, English
Literature as a 'subject' has been founded upon a series of
uncertainties and conflicts. Indeed it can be and has been
argued that for a discipline professing *criticism* a sense of
crisis is an appropriate, even favourable, condition. Leaving
aside this general consideration, what is new in the recent
disputes over the study of literature is the existence of a
growing 'opposition' movement which has begun to question
some of the long-standing assumptions of traditional literary
criticism embedded in the very title 'English Literature':
both the status of Great (and hence capitalized) Literature
and the Englishness of the subject's concerns and methods.
One of the major contentions of this opposition is that
traditional English Studies are 'ideological' − a claim which
is often greeted with some scepticism, for two related reasons:
one a problem of terminology and the other a problem of
approach.
   In the first place it is objected that nothing could be more
harmless, apolitical, and undogmatic than the study and
criticism of literary works; surely it is the femin-, Marx-,
structural-, and other -ists who are 'ideological' in importing
their doctrines into this neutral area (not to say paranoid in
typically detecting conspiracies where none exist)? The
objection employs a particular sense of the term 'ideology',
first used by Napoleon Bonaparte to brand his democratic
political opponents as impractical philosophical dogmatists.
This sense was revived by American political 'theorists' in

the Cold War for much the same purposes and today enjoys currency in journalistic usage; it refers to a conscious and explicit theory adopted by a minority. On the other hand the Marxist sense of the term adopted by the 'opposition' in literary studies has a wider and more complex reference, denoting those usually unspoken assumptions upon which the most untheoretical, undogmatic, and 'common-sense' arguments rest; in particular the assumption that the existing institutions and values of society are natural and eternal rather than artificial and temporary.[1]

This understanding of ideology, incidentally, has no need whatever (in contrast with the Bonapartist usage) to invent conspiracies, since it sees ideologies simply as the line of least resistance taken in interpreting existing circumstances; as 'lazy' reflections of the world around them, which either do not bother or do not want to consider the evidence unfavourable to their implicit tenets. It is this charge (to simplify drastically) which is being levelled at conventional literary study; the fact that the guardians of traditional approaches do not like to regard themselves as -ists of any kind does not affect its validity.

The second problem, which compounds the unhappy incompatibility of terms, follows from the nature of ideologies according to this wider conception. For if the major common factor of ideologies is their assumption that the existing order of things (whether it be 'femininity', private property, or the artistic portrayal of 'human nature') originates in nature and carries on into eternity, then as far as the theoretical refutation of such an assumption goes — which is well short of its (to borrow a phrase) practical criticism — a major if not predominant role in the counter-argument must be allotted to historical example. Exposure of logical inconsistencies in an ideological argument can go a certain distance, but it is only history which can challenge any assumption of 'timelessness' at its root. The irony of the recent debates is that those most eager to argue that English studies are ideological are often those whose adopted methods are precisely the least equipped to establish such a contention. There has been much that is valuable and stimulating in the belated absorption of structuralism into English studies in

Britain; yet its paralysing anti-historical tendencies have severely blunted the very critique which many of the structuralists wish to make of traditional 'English Literature'. Without recourse to history, in short, the contention that this traditional model is ideological will remain unconvincing.[2]

The purpose of this book is in part to redress that shortcoming within the current questioning of 'English Literature', and within the attention of its students generally. I have found in discussing my preparation of this work with undergraduate students of English in every case an astonishment at the comparative novelty of their chosen subject within the history of higher education. It would seem that the study of English Literature is accepted by most of its practitioners as a 'natural' activity without an identifiable historical genesis. With some qualification, the same goes for the discourse — literary criticism — which dominates the subject. It is perhaps generally known, for example, that Eliot said there was a dissociation of sensibility in the seventeenth century and that Arnold and Leavis claimed in some way a moral purpose for literature; but just why they should have been moved to say such things (and it was certainly not in order to provide topics for examination papers) is a question which seems rarely to be asked within a discipline so unconcerned to examine its own history. Even leaving aside the very favourable conditions which it provides for ideological assumptions of timelessness and naturalness, such a state of affairs is in itself an unhealthy one for literary study, fostering within it a passive and indeed uncritical attitude. With these considerations in mind, my approach has been a deliberately unsophisticated attempt to drag back into the light the views taken by the founders of modern English Studies and literary criticism regarding the wider social effects and aims of this activity; to restore to what is now a severely truncated vision of criticism's recent past those neglected but essential statements of its original purpose as an active participant in society.

Accordingly, the following chapters will be concerned with the development of certain ideas of literary criticism's social function in an important period of English criticism, attempting in particular to trace the contribution of Matthew

Arnold to the literary-critical renaissance of the 1920s and early 1930s represented by the writings of T. S. Eliot, I. A. Richards, and F. R. and Q. D. Leavis. The achievement of these critics will be examined as part of a common development of the ideal of 'practical criticism' — to be understood here in a sense wider than that of the technical exercise to which the phrase usually refers, denoting rather a 'practicality' in which criticism seeks a real practical effect upon society, directly or indirectly. First, the prose writings of Matthew Arnold will be surveyed as a consistent unit including not only strictly literary but also theological, political, and educational works, illustrating the extent to which Arnold expanded the duties of literary criticism into these areas, and how, in turn, his conception of society transformed his vision of the function of criticism. After a brief examination of Walter Pater's development of Arnoldian ideas, some of the early arguments for the educational importance of the study of English Literature will be reviewed in Chapter 3, in particular those emphasizing the possibilities of this subject as a civilizing and humanizing agency of beneficial social consequences; and some of the initial problems of the subject at university level will be touched on. Chapter 4 will be devoted to the impetus given to these arguments by the war of 1914–18, and the promotion of English as a study conducive to national pride and unity. Chapters 5, 6, and 7 will examine in turn the early critical writings of T. S. Eliot, I. A. Richards, and the Leavises, attempting to show how they revived and modified in different ways the work of Arnold, in 'practical' directions. Finally, the concluding chapter will review this line of thinking on the importance of literary criticism, itemizing the common characteristics of the writers examined and the senses in which they made criticism 'practical'. This introductory chapter aims first to set the object of our investigation within the context of the history of criticism.

Discussion of criticism is apt to be considered a rather introspective diversion. But it is a fact too often forgotten that the real content of the school and college subject which goes under the name 'English Literature' is not literature in the primary sense, but *criticism*. Every school student in

British education is required to compose, not tragic dramas, but essays in criticism. The historical study of criticism is needed at least to explain this now widespread practice; and it deserves to be defended further against charges of parasitism: as if criticism itself were not sufficiently removed from its source and occasion in literary works, why tolerate a discourse set apart at two (and in parts of this chapter, for example, even at three) removes from the real object of literary studies? Charges of parasitism begin from the assumption — very easily made, for a number of reasons — that literary criticism as an activity owes its entire existence to the literary works upon which, like an ungrateful child, it presumes to pass comment. And in English literature from the Restoration onwards, there is a whole tradition of bitter jibes by authors against critics to endorse and win sympathy for this scale of priority: Fielding's description in *Tom Jones,* for example, of critics as reptiles, or Byron's allegation that it was the critics who killed John Keats. In one extremely hypothetical sense this assumption may be true, in that if there were no Literature there would be no criticism — and even this remote case is doubtful (if Literature did not exist, it can be argued, it would have to be invented — a task for criticism). Yet literary history shows that criticism does not 'shadow' some primary literary progenitor in any such simple fashion. In the first place, the recurrent case of authors neglected in their own time but acclaimed decades or centuries later is only the most noticeable aspect of a reverse process by which criticism 'creates' what is accepted as Literature. And furthermore, there are important instances of major lines of literary work failing to 'produce' a critical offspring at all for very long periods: the English novel, most remarkably, went through its classical period in the nineteenth century to no significant critical accompaniment, while Shakespeare and Wordsworth were abundantly discussed all the while. Such time-lags indicate that the history of literary criticism moves according to laws to a great degree autonomous from those of its alleged parent, that the productive and receptive sides of literary history do not make anything like a perfect fit, and therefore that other factors apart from literary works themselves go into the making of criticism. These are the considerations which justify the historiography of criticism as a distinct study, the third 'level' (if the hierarchy still stands) accounting, among other things, for the discrepancy between the first two.

Unfortunately most histories of criticism adopt the 'parasitic' view in which criticism emerges one-dimensionally from primary literary sources. The problem was well put by George Watson in his book *The Literary Critics*, which represents the first sustained attempt to break out of it:

But all previous histories — Saintsbury and Atkins in their day, as much as Wellek and Wimsatt in ours — have assumed that what we call literary criticism is, with some embarrassing exceptions, a single activity, and that its history is the story of successive critics offering different answers to the same questions. We may call this the Tidy School of critical history.[3]

Watson offers instead 'a record of chaos marked by sudden revolution',[4] and a specification of criticism according to three types, Legislative, Theoretical, and Descriptive; his own account covering only the third category. A thorough history of criticism would need to go still further in refusing the 'tidy' method by recognizing among other things that critics and periods of criticism differ in their degrees of critical self-consciousness. Far from restricting itself to a naive 'description', the general tendency in English criticism since Dryden has been towards a heightened self-consciousness in the major critics (or, put another way, it is the most critically self-conscious writers who stand out as the major critics). Criticism passes from the subordinate position of defender of poetry to a position of self-appointed authority from which it can turn to the offensive, in social as well as literary comment: since Arnold in particular, English criticism has, as Patrick Parrinder has put it, lost its innocence.[5]

The kind of distinction I have in mind is best exemplified in one of the shortest but still most stimulating histories of criticism: T. S. Eliot's *The Use of Poetry and the Use of Criticism* (1933), which adopts the important proposition that 'our criticism, from age to age, will reflect the things that the age demands'.[6] Eliot rejects the view, similar to the 'anti-parasitical' arguments discussed above, that criticism is a symptom of decadence, and attempts instead to explain criticism's emergence historically as a necessary adjustment to changed conditions of authorship. Rather than simply relate the successive 'contributions' of critics, he sets out to ask why, for example, Dryden and Johnson should want to write about poetry and poets. From such questions, Eliot is able to proceed to the problem of what we have referred to as the relative 'self-consciousness' of critics. He points out, for example, the advantages enjoyed by Samuel Johnson: 'Had he lived a generation later, he would have been obliged to look more deeply into the foundations, and so would have been unable to leave us an example of what criticism ought to be for a civilisation which, being settled, has no need, while it lasts, to enquire into the function of its parts.'[7] By contrast with this confidently 'naive' criticism, modern criticism has to function in a world where large assumptions can no longer be relied upon:

...when the poet finds himself in an age in which there is no intellectual aristocracy, when power is in the hands of a class so democratised that whilst still a class it represents itself to be the whole nation; when the only alternatives seem to be to talk to a coterie or to soliloquise, the difficulties of the poet and the necessity of criticism become greater.[8]

It is less the supposedly eternal questions invited by poetry than the problems posed by society which, for Eliot, determine the development of criticism: he insists, for example, that Wordsworth's statements on poetic diction were really animated by his social concerns. Briefly, his argument is 'that the development and change of poetry and of the criticism of it is due to elements which enter from outside'.[9] Acceptance of this argument does away with the usual conception of criticism as a pre-given set of questions which reproduces itself under its own internal momentum; the importance of 'outside' factors can be recognized.

The very vocabulary of literary criticism, its constitutive metaphors, ought to be sufficient to betray the pressure of such outside factors. Two frames of reference in particular recur almost monotonously in critical discourse: the judicial or forensic, and the economic. 'Judgement' and 'evaluation' are the two terms most commonly resorted to by critics to define their task, and the clusters of metaphors which they carry with them — of courts and tribunals, of value and debasement — are not at all arbitrary. They register real sources from which criticism derives, 'from outside', its status; real forces which impinge upon the production and reception of literary works.

In the first place the 'judgement' of literary works has a real extra-metaphorical equivalent in the fact that these works have always endured a degree of censorship and legal restraint upon their publication and dissemination. In casting themselves as 'judges' or as witnesses for the defence, critics habitually mimic the authority of more powerful assessors of literature. In some cases, indeed, opinions of a distinctly literary nature may become incorporated directly into forensic procedure: notable cases include Oscar Wilde's defence of his letter to Lord Alfred Douglas, from the dock at his sodomy trial, on the grounds that it was a 'prose poem';[10] or Brecht's wrangles with the House Committee on Un-American Activities over the correct translation and meaning of his poems 'In Praise of Learning' and 'Forward, We've Not Forgotten'.[11] Still more impressive as an example of criticism's forensic figures coming true is the 'Lady Chatterley' trial (*Regina* v. *Penguin Books Limited*, 1960), a test case under the new Obscene Publications Act (1959) which required that expert witnesses be called to judge a book's literary merit. It was the long list of literary-critical 'expert witnesses' which persuaded the jury to acquit, while the prosecution found itself arguing on grounds associated with the literary-critical concept of the 'intentional fallacy'.[12]

Although its real effect was to relax censorship, under this law adverse pronouncements by literary critics can lead directly to the suppression of a literary work. These are of course very special cases, but they highlight a normally submerged component present in much critical discourse. Under normal circumstances the 'verdict' of criticism will have extra-legal consequences only, for example for the policy of publishing houses, but consequences which may be effectively identical. Despite the silence on this point in the theories and histories of criticism, there is no impassable gulf between censorship and criticism; the former may often be seen as a paradigm of the latter or, so to speak, its armed wing. In this light, George Watson's assertion that criticism 'presupposes an open society'[13] appears as an extremely doubtful piece of Cold War sophistry. For example, it would take considerable special pleading to demonstrate that the society in which Coleridge, not long before the Peterloo massacre, published his *Biographia Literaria* was a distinctly 'open' one; or that because of his compromises with Stalinism Georg Lukács was not a literary critic. Criticism from Plato onwards has, on the contrary, presupposed censorship, banishment, and official persecution in the very language of its 'judgements' and in its images of its own authority.

The second constitutive metaphor upon which criticism has drawn is more suitable to civil society than to the workings of state surveillance. The vocabulary of 'value' achieves particular prominence in criticism when critics become advisers to a class of literary consumers anxious to know the worth of their purchases. In this more or less free market of literary commodities, the staple critical genre becomes the book review, and its equivalent, the publisher's 'blurb' — again an area too sordid and worldly to have been treated by the theories and histories of criticism; it is not generally counted among T. S. Eliot's literary achievements, for example, that he was Faber's best 'blurb'-writer,[14] nor could these texts be considered part of the Eliot canon under a critical orthodoxy in which advertising is considered untouchable. The literary vocabulary of value is, again, not an arbitrary figure of speech but the mark upon criticism of considerations which no book reviewer can altogether ignore, absorbed into criticism 'from outside'.

These observations are made to enforce the point that literary criticism is not a discourse born fully armed from the head of, say, Aristophanes or Plato, but a composite discourse. Even (indeed, especially) in those major deceptive instances — Dryden, Coleridge, Arnold, Eliot — where important critical figures have also been important poets, criticism does not derive from poetry, or even solely from the technical framework of rhetoric (itself, of course, contiguous with forensic and political traditions), but carries significantly more than just traces of other discourses, notably the economic, political, and judicial. It will be worth briefly reviewing

some instances. Plato, sometimes taken as the starting-point
for Western criticism, presents almost an extreme confirmation
of this view, his comments on poetry in *The Republic* leading
up to his political decision in Book 10 to expel the poets
from his ideal state. To turn to England, Dryden's status as
'Father' of English criticism owes much to his particular fusion
of political and poetical interests into a new kind of critical
outlook. Of his major critical work, *Of Dramatic Poesy: An
Essay* (1668), Dryden wrote that its purpose was 'chiefly to
vindicate the honour of our English writers, from the censure
of those who unjustly prefer the French before them',[15] and
it is not by accident that its dialogues are set against the
background noise of an English naval victory. Later critics
of the 'neo-classical' period constantly play upon the contrast
between the regularity insisted upon by post-Restoration
taste and the barbarity both of literature and of politics in
the Civil War and before. Later, Wordsworth's challenging
of the special poetic diction encouraged by neo-classical
tastes was itself informed partly by political motives; an
attempt to 'democratize' the reading public's attitude to
poetic language, in turn challenged by Coleridge's insistence
upon a hierarchy of discourses and mental faculties.

The cases briefly noticed above should give some indication
of the extent to which literary criticisms have incorporated
'from outside' various other elements, from patriotism to
more elaborate political and social ideologies. A further
development or exaggeration of this factor in the history
of criticism, however, takes place when its use becomes
conscious and deliberate, when critics conceive for themselves
a social function more extensive than simply the defence of
the national literary heritage against foreign competition. In
English criticism, such a transformation is effected most
decisively by the work of Matthew Arnold — to be examined
in the next chapter. This work is itself partly informed by an
important precedent which should be noted by way of
introduction to the theme of criticism's social function in
the modern English tradition: the writings of the French
literary critic Charles-Augustin Sainte-Beuve.

Sainte-Beuve's critical work, in particular the enormous
production represented in his *Causeries du lundi* (1849–69),

was the most important critical 'consolidation' of the nine-teenth century, comparable with that carried out more concisely by Samuel Johnson. Arnold had enormous admiration for Sainte-Beuve as a critical authority,[16] using him as a model of what a modern critic should be. He learned, for example, from Sainte-Beuve's attack on the excesses of Romantic literature:

Une terrible émulation et comme un concours furieux s'était engagé dans ces dernières années entre les hommes les plus vigoureux de cette littérature active, dévorante, inflammatoire. Le mode de publication en feuilletons, qui obligeait, à chaque nouveau chapitre, de frapper un grand coup sur le lecteur, avait poussé les effets et les tons du roman à un diapason extrême, désespérant, et plus longtemps insoutenable. Remettons-nous un peu. En admirant le parti qu'ont su tirer souvent d'eux-mêmes des hommes dont le talent a manqué des conditions nécessaires à un développement meilleur, souhaitons à l'avenir de notre société des tableaux non moins vastes, mais plus apaisés, plus consolants, et à ceux qui les peindront, une vie plus calmante et des inspirations non pas plus fines, mais plus adoucies, plus sainement naturelles et plus sereines.[17]

(A dreadful rivalry, a sort of furious contest, has begun in recent years among the strongest men in this potent, all-consuming, and inflammatory literature. The serial mode of publication, which demands that each new chapter should stun the reader, has forced the tone and feeling of the novel to a desperate and eventually unsustainable extremity of pitch. Let us calm down a little. While admiring that school which has often been able to foster men whose talent has lacked the necessary conditions in which to flourish, we wish, for the future of our society, for pictures just as extensive but calmer and more consoling, and for those who paint them, a more soothing life, and inspirations just as fine but more softened, more sanely natural, and more serene.)

It is in exactly these terms — sanity, consolation, serenity — that Arnold was to conduct his own critique of Romanticism a few years later; and it is the same fear of inflammatory extremism which guided Arnold's writings on social and political as much as on literary questions. The continuity between the two realms, between political order and literary order, is emphasized particularly in Sainte-Beuve's early *Causeries*, where he elaborates on his idea of the restraining role of the literary critic in accomplishing a cultural 'restoration' alongside the restoration of political authority after revolutionary upheavals (the case of Dryden comes strikingly to mind):

En 1800, on était à l'une de ces époques où l'esprit public tend à se reformer. Il y avait lutte encore, mais aussi, déjà, ensemble et concert; il y avait lieu à direction. On sortait d'une affreuse et longue période de licence, de dévergondage et de confusion. Un homme puissant replaçait sur ses bases l'ordre social et politique. Toutes les fois qu'après un long bouleversement l'ordre politique se répare et reprend sa marche régulière, l'ordre littéraire tend à se mettre en accord et à suivre de son mieux. La critique (quand critique il y a), à l'abri d'un pouvoir tutélaire, accomplit son oeuvre et sert la restauration commune. Sous Henri IV, après la Ligue, on eut Malherbe; sous Louis XIV, après la Fronde, on eut Boileau. En 1800, après le Directoire et sous le premier Consul, on eut en critique littéraire le monnaie de Malherbe et de Boileau, c'est-à-dire des gens d'esprit et de sens, judicieux, instruits, plus ou moins mordants, qui se groupèrent et s'entendirent, qui remirent le bon ordre dans les choses de l'esprit et firent la police des Lettres. Quelques-uns firent cette police fort honnêtement, d'autres moins; la plupart y apportèrent une certaine passion, mais presque tous, à les prendre au point de départ, agirent utilement.

A ces époques qui suivent un grand danger et où l'on vient d'échapper à des grands malheurs, on sent très distinctement le bien et le mal en toutes choses; on est disposé à exclure, à interdire ce qui a nui, et c'est le moment où le critique trouve le plus d'appui et de *collaboration* dans le public.[18]

(In 1800 we were in one of those periods when public opinion rights itself. There was struggle, but already there was cohesion and co-operation as well; there was room for leadership. We were emerging from a long and terrible period of licentiousness, profligacy, and confusion. A strong man was re-establishing social and political order. Every time political order restores itself and resumes its regular course, literary order tends to fall into line and follow suit as best it can. Under the wing of a guardian power, criticism (when there is criticism) does its work, and serves this combined restoration. Under Henri IV, after the Ligue, we had Malherbe; under Louis XIV, after the Fronde, we had Boileau. In 1800, after the Directory and under the first Consul, we had in literary criticism the small change from the currency of Malherbe and Boileau; that is to say, intelligent people of sound understanding, judicious, learned, and more or less acute, who gathered together in agreement, who re-established order in intellectual life and acted as a literary police. Some carried out this policing very honourably, some less honourably; most brought a certain passion to this work, but nearly all, to begin with, acted usefully.

In these periods which follow a great danger, when great evils have just been escaped, the good and the bad in all things are felt very distinctly; we are inclined to exclude, to prohibit whatever has done us harm; and it is at these times that the critic finds the most support and *collaboration* from the public.)

Sainte-Beuve's preoccupation with this function of literary criticism arose from the recognition that he himself was living through just such a dangerous period, following the revolutionary turmoil of 1848. The political 'strong man' for whom Sainte-Beuve was to be the literary equivalent was Louis Bonaparte, and it was in the (far from 'disinterested') Bonapartist journal *Le Constitutionnel* and the official government paper *Le Moniteur* that he published his *Causeries*. The restoration of authority, both political and literary, was a combined necessity for his own time, as he insisted in his *Causerie* on Boileau, referring to the healthy influence of Boileau and of Louis XIV on Racine, La Fontaine, and Molière:

Boileau, c'est-à-dire le bon sens du poëte critique, autorisé et doublé de celui d'un grand roi, les contint tous et les contraignit, par sa présence respectée, à leurs meilleures et à leurs plus graves oeuvres. Savez-vous ce qui, de nos jours, a manqué à nos poëtes, si pleins à leur début de facultés naturelles, de promesses et d'inspirations heureuses? Il a manqué un Boileau et un monarque éclairé, l'un des deux appuyant et consacrant l'autre. Aussi ces hommes de talent, se sentant dans un siècle d'anarchie et d'indiscipline, se sont vite conduits à l'avenant; ils se sont conduits, au pied de la lettre, non comme des nobles génies ni comme des hommes, mais comme des écoliers en vacances. Nous avons vu le résultat.[19]

(Boileau, that is to say the sound judgement of the poet and critic, authorized and reinforced by that of a great king, restrained them all by his respected presence and constrained them to produce their best and their most serious works. Do you know what in our age has been lacking in our poets, who started out so naturally gifted, so full of promise and felicitous inspiration? They lacked a Boileau and an enlightened monarch, each supporting and sanctioning the other. And so these talented men, feeling themselves to be in a century of anarchy and indiscipline, have behaved accordingly; they have behaved not like noble spirits, nor like men, but literally like schoolboys on holiday. We have seen the result.)

There is no great distance between such observations and Arnold's opposition between Culture and Anarchy — whether the anarchy of Romanticism on the bookshelves or of demonstrations on the streets — and his recourse to the authority of the state as a higher agency of restraint. Closer still to Arnold's aims is the conception offered in another of Sainte-Beuve's *Causeries* of a higher inspiration required for the preservation of civil peace:

[Paul-Louis Courier] oublie trop que *Georges le laboureur, André le vigneron, Jacques le bonhomme* (comme il les appelle) n'ont rien qui les élève et les moralise, qui les détache de ces intérêts privés auxquels ils sont tous acharnés et assujettis; qu'à un moment donné, s'il faut un effort, un dévouement, une raison supérieure d'agir, ils ne la trouveront pas, et qu'à telles gens il faut une religion politique, un souvenir ou une espérance qui soit comme l'âme de la nation, quelque chose qui, sous Henri IV, s'appelait le Roi, qui plus tard s'appellera l'Empereur, qui, dans l'avenir, sera je ne sais quel nom: sans quoi, à l'heure du péril, l'esprit d'union et d'unité, le mot d'ordre fera faute et la masse ne se soulèvera pas.[20]

([Paul-Louis Courier] is too inclined to forget that 'Georges the ploughman, André the wine-grower, and Jacques the little man' (as he calls them) have nothing which can lift them or raise their moral standards, which can draw them from their private interests, to which they are all stubbornly attached; that if what is needed at a given moment is an effort, a self-sacrifice, a higher motive, they will not find it, and that for these people a political religion is required, a memory or a hope which could be the soul of the nation, something which under Henri IV was called the King, which will later be called the Emperor, and whose name in the future is uncertain; without which, in the hour of danger, the spirit of concord and unity will lack a watchword, and the masses will not rise.)

An influence capable of elevating and 'moralizing' the masses, to replace the old monarchical and imperial ideals, but as yet undefined: it was the search for this which Sainte-Beuve left to his English disciple, with strong enough hints that the 'literary order' would form part of it. Arnold himself had followed the political crisis of 1848 from afar, and his observations bear on the search. While admiring the intelligence of the French masses, he was afraid that their restlessness would infect their more primitive British counterparts, as he confided to his sister at the time.[21] The dynamics of mass movements concerned Arnold greatly at this time; and in the middle of the French crisis, he came to conclusions on this question which he expressed to his friend Arthur Clough: 'Seditious songs have nourished the F[renc]h people much more than the Socialist 'philosophers' though they may formalize their wants through the mouths of these.'[22] By seeking the sources of the French popular movement's 'nourishment' in its cultural rather than political or philosophical features, Arnold was already beginning to formulate what Sainte-Beuve, under the shadow of the same disturbing events, was to seek as the missing inspiration required for national unity and reconciliation. A new vocabulary of 'Culture' was on its way to transforming the scope and responsibilities of English literary criticism.

An introductory survey of the 'social' component in critical history has been necessary in order to emphasize the importance of examining the ideas and assumptions of literary critics concerning the social effects and functions of their work. It is insufficient, especially after Arnold, to regard literary critics' social, political, or religious interests as separate pursuits outside their literary criticism 'proper'. A critical history which adopted such a separation would be inadequate, if only because it would involve ignoring what the acknowledged leaders of English critical thought — Arnold, Eliot,

Richards, the Leavises — did after all say about the purpose of their work.

This book does not attempt a 'rounded' study of the critics discussed: it leaves aside many questions of their technique and of their particular judgements of authors and literary works, and in the case of the twentieth-century critics it stops well short of discussing their full careers. Still less is it a chronological chapter of the more thorough kind of critical history envisaged above. It is instead necessarily selective and partial, concentrating on major figures and developments, and seeking to follow a particular 'line' in the development of criticism rather than show the available material in all its variety. The study draws to a close, chronologically, with the launch of *Scrutiny* in 1932[23] — not an arbitrary point at which to end, but a point at which the issue of criticism's social function takes on a different, more organized form, and when other possible developments of this tradition dispersed.

NOTES

1. See Raymond Williams, *Keywords: A Vocabulary of Culture and Society* (1976), 126-30; and more fully, Jorge Larrain, *The Concept of Ideology* (1979).
2. A belated recognition of this is registered in some contributions to the recent collection *Re-Reading English* edited by Peter Widdowson (1982).
3. George Watson, *The Literary Critics* (1962; 2nd edn., Harmondsworth, 1973), 2.
4. Ibid., 3.
5. Patrick Parrinder, *Authors and Authority: A Study of English Literary Criticism and its Relation to Culture 1750-1900* (1977), 1.
6. T. S. Eliot, *The Use of Poetry and the Use of Criticism* (1933), 141.
7. Ibid., 65
8. Ibid., 22.
9. Ibid., 127.
10. H. Montgomery Hyde, *The Trials of Oscar Wilde* (1948; NY, 1973), 244-6.
11. See Eric Bentley (ed.), *Thirty Years of Treason: Excerpts from Hearings before the House Committee on Un-American Activities 1938-1968* (1972), 217-20.

12. C. H. Rolph (ed.), *The Trial of Lady Chatterley: Regina v. Penguin Books Limited* (Harmondsworth, 1961), 7, 219-20.

13. Watson, *Literary Critics*, 1.

14. See F. V. Morley, 'T. S. Eliot as a Publisher', in *T. S. Eliot, A Symposium*, ed. Richard March and Tambimuttu (1948), 60-70.

15. John Dryden, *Selected Criticism*, ed. James Kinsley and George Parfitt (1970), 19.

16. See Arnold Whitridge, 'Matthew Arnold and Sainte-Beuve', *PMLA*, liii (1938), 303-13; and R. H. Super, 'Documents in the Matthew Arnold — Sainte-Beuve Relationship', *Modern Philology* lx (1963), 206-10.

17. C. -A. Sainte-Beuve, *Causeries du lundi*, ii. 463 ('Balzac', 2 September 1850). The translation which follows, here and below, is my own.

18. Ibid., i. 373-4 ('M. de Feletz, et de la Critique Littéraire sous l'Empire', 25 February 1850).

19. Ibid., vi. 511-12 ('Boileau', 27 September 1852).

20. Ibid., vi. 347 ('Paul-Louis Courier', 21 July 1852).

21. 'What agitates me is this, if the new state of things succeeds in France, social changes are *inevitable* here and elsewhere ... but, without waiting for the result, the spectacle of France is likely to breed great agitation here, and such is the state of our masses that their movements now *can* only be brutal plundering and destroying.' *Letters of Matthew Arnold 1848-1888*, ed. George W. E. Russell (2 vols., 1895), i. 6 (10 March 1848).

22. *The Letters of Matthew Arnold to Arthur Hugh Clough*, ed. H. F. Lowry (1932), 74 (8 March 1848).

23. Readers seeking an account of developments after 1932 are strongly recommended to consult Francis Mulhern's *The Moment of 'Scrutiny'* (1979).

2. MATTHEW ARNOLD'S INNOCENT LANGUAGE

> There be who perpetually complain of schisms and
> sects . . . They are the troublers, they are the dividers
> of unity, who neglect and permit not others to unite
> those dissever'd peeces which are yet wanting to the
> body of truth.
>
> Milton, *Areopagitica*

i. *The future of poetry and the function of criticism*

Matthew Arnold is rightly acknowledged as the founder of
a distinctly 'modern' movement in English literary criticism;
not for any specific judgement upon authors or poems but for
his emphatic readjustment of literary-critical discussion
towards questions of literature's social function, and consequently the social function of criticism itself. In one short
essay, 'The Study of Poetry' (1880), he gathered the high
claims made for poetry by his Romantic predecessors and
reformulated them in something approaching a prophecy of
the coming superiority of poetry over entire areas of human
thought and activity:

The future of poetry is immense, because in poetry, where it is worthy
of its high destinies, our race, as time goes on, will find an ever surer and
surer stay. There is not a creed which is not shaken, not an accredited
dogma which is not shown to be questionable, not a received tradition
which does not threaten to dissolve. Our religion has materialised itself
in the fact, in the supposed fact; it has attached its emotion to the fact,
and now the fact is failing it. But for poetry the idea is everything; the
rest is a world of illusion, of divine illusion. . . . The strongest part of
our religion today is its unconscious poetry.

We should conceive of poetry worthily, and more highly than it has
been the custom to conceive of it. We should conceive of it as capable
of higher uses, and called to higher destinies than those which in general
men have assigned to it hitherto. More and more mankind will discover
that we have to turn to poetry to interpret life for us, to console us,
to sustain us. Without poetry, our science will appear incomplete; and
most of what now passes with us for religion and philosophy will be

replaced by poetry. Science, I say, will appear incomplete without it. For finely and truly does Wordsworth call poetry 'the impassioned expression which is in the countenance of all science'; and what is a countenance without its expression? Again, Wordsworth finely and truly calls poetry 'the breath and finer spirit of all knowledge': our religion, parading evidences such as those on which the popular mind relies now; our philosophy, pluming itself on its reasonings about causation and finite and infinite being; what are they but the shadows and dreams and false shows of knowledge? The day will come when we shall wonder at ourselves for having trusted to them, for having taken them seriously; and the more we perceive their hollowness, the more we shall prize 'the breath and finer spirit of knowledge' offered to us by poetry.

But if we conceive thus highly of the destinies of poetry, we must also set our standard for poetry high, since poetry, to be capable of fulfilling such high destinies, must be poetry of a high order of excellence. We must accustom ourselves to a high standard and to a strict judgement. [*CPW* ix. 161-2.]

Here, at its most challenging and controversial, is Arnold's declaration of faith. It marks a transformation in English criticism, from the defence of poetry to a bold offensive against poetry's potential competitors: religion, philosophy, science. At the same time it assigns to literary criticism responsibilities no less awesome than those of poetry, introducing into English critical writing a new sense of self-consciousness, a new sensitivity to the wider social and cultural duties befitting its special guardianship.

In attempting to understand his contribution to criticism, it is necessary at once to notice how Arnold addresses himself in 'The Study of Poetry' to poetry's future. A definite historical perspective is involved in his advocacy of poetry: a conviction that the time is ripening for poetry to take up its rightful place in human affairs even if it now lies neglected. Poetry's future was to be immense, but its present state was troubled. The present time was, for Arnold, an unpoetical age, a depressingly difficult interim period of transition in which obstacles had to be cleared and the ground laid for poetry's future. Both the conviction of criticism's enormous social responsibilities, and the specific historical pattern which laid such responsibilities upon it, are the themes concentrated in the title of Arnold's other major manifesto, 'The Function of Criticism at the Present Time' (1864). If 'The Study of Poetry' was Arnold's manifesto for the immense future of poetry, the earlier essay was the programme for the present, slow, and apparently unrewarding preparatory tasks of criticism. Here, Arnold argues that a golden age of creative literature equivalent to those of Aeschylus or Shakespeare is not an immediate possibility for his generation, but a 'promised land, towards which criticism can only beckon' [*CPW* iii. 285]. Such a creative age would require not just individual poetic genius, but the availability of a current of fresh ideas upon which poets could draw. Poets need such data or raw materials for their work, and cannot press ahead without them:

. . . the exercise of the creative power in the production of great works of literature or art, however high this exercise of it may rank, is not at all epochs and under all conditions possible; and . . . therefore labour may be vainly spent in attempting it, which might with more fruit be used in preparing for it, in rendering it possible. This creative power works with elements, with materials; what if it has not those materials, those elements, ready for its use? In that case it must surely wait till they are ready. [*CPW* iii. 260.]

While the creative power will have to wait, the critical power has its work to do, the task of seeking out the best materials in all branches of intellectual activity. The poets themselves cannot be expected to discover and circulate their own new ideas, 'for creative literary genius does not principally show itself in discovering new ideas, that is rather the business of the philosopher. The grand work of literary genius is a work of synthesis and exposition, not of analysis and discovery.' [*CPW* iii. 260-1.]

Arnold argues for a distinction to be made, which is at once a division of labour between the thinker and the poet (the critic playing the intermediary role), and a chronological division between the period of accumulation of raw materials and the period of creative production. 'Criticism first; a time of true creative activity, perhaps, – which, as I have said, must inevitably be preceded amongst us by a time of criticism, – hereafter, when criticism has done its work.' [*CPW* iii. 269.] By this gesture of postponement (a recurrent principle in his work) Arnold claimed for criticism in a very particular sense a place in history. While creation waits for its appropriate moment in the hereafter, criticism is allotted a predominant role in the present, as the supervisor and sifter of all intellectual production. But this privileged position is not one which can be secured or retained easily, having to be defended, indeed, against an array of enemies.

In seeking a special place in history for criticism, Arnold found himself obliged immediately to illustrate his idea of history with reference to both cultural and political events. To advocate a postponement of creative endeavour in favour of strictly preparatory work meant that some explanation of the importance of historical 'preparedness' and 'unpreparedness' was called for. Arnold's first victim in 'The Function of Criticism at the Present Time' was the English poetry of the first quarter of the nineteenth century. For all its creative force and energy, it 'had about it in fact something premature', it 'did not know enough', lacking as it did a thriving intellectual culture around it from which to draw ideas [*CPW* iii. 262]. The second example of a movement flourishing too suddenly to sustain itself was the French Revolution, which Arnold's hindsight recognizes as 'the greatest, the most animating event in history'; although it penetrated a whole nation with an enthusiasm for the laws of reason, the Revolution was regrettable in its fatal 'mania for giving an immediate political and practical application to all these fine ideas of the reason' [*CPW* iii. 265]. Like English Romantic poetry, the French Revolution was premature; it had thrust itself forward before its ideas had sufficiently matured. While explaining this conception, Arnold not only provides a succinct statement of his political philosophy, but adds to the two major distinctions made in the essay — between critical and creative periods, and between thinkers and poets — a third distinction: between the sphere of ideas and the sphere of politics and practice.

Ideas cannot be too much prized in and for themselves, cannot be too much lived with; but to transport them abruptly into the world of politics and practice, violently to revolutionize the world to their bidding, — that is quite another thing. . . . *Force till right is ready;* and till right is ready, force, the existing order of things, is justified, is the legitimate ruler. But right is something moral, and implies inward recognition, free assent of the will; we are not ready for right, — *right*, so far as we are concerned, *is not ready*, — until we have attained this sense of seeing it and willing it. The way in which for us it may change and transform force, the existing order of things, and become, in its turn, the legitimate ruler of the world, should depend on the way in which, when our time comes, we see it and will it. Therefore, for other people enamoured of their newly discerned right, to attempt to impose it on us as ours, and violently to substitute their right for our force, is an act of tyranny, and to be resisted. It sets at nought the second great half of our maxim, *force till right is ready*. This was the grand error of the French Revolution; and its movement of ideas, by quitting the intellectual sphere and rushing furiously into the political sphere, ran, indeed, a prodigious and memorable course . . . [*CPW* iii. 265-6.]

Arnold's contribution to criticism is inseparable from his conception of history and of the great political movements of his time. The linked historical principles, as Arnold saw them, of prematurity and postponement originate from his assessment of early nineteenth-century history and its likely consequences. These principles, which require the partitioning of creation and criticism, of poet and thinker, and of practice and ideas in Arnold's cultural campaigns, form the basis of his whole critical project. Criticism, for him, involved a long-term programme for the reform of Britain's entire intellectual life, an effort to temper and soften the stridency of contemporary political and religious partisanship, a strategy for containing radical new movements within traditional frameworks in the interests of social and cultural harmony; a stance summed up in the term borrowed from Sainte-Beuve — 'disinterestedness'. At the same time as he postpones poetic creation to the hereafter, Arnold imposes an important restriction upon its preparatory agent, criticism: the critic must set an example by abstaining from all practical political quarrels, all the better to influence, in the long run, the thought and eventually the practice of society. Such a stance is proposed as a conscious rejection of the habits of previous decades, and as an investment for decades to come:

Everything was long seen, by the young and ardent amongst us, in inseparable connection with politics and practical life. We have pretty well exhausted the benefits of seeing things in this connection, we have got all that can be got by so seeing them. Let us try a more disinterested mode of seeing them; let us betake ourselves to the serener life of the mind and spirit. This life, too, may have its excesses and dangers; but they are not for us at present. Let us think of quietly enlarging our stock of true and fresh ideas, and not, as soon as we get an idea, be running out with it into the street, and trying to make it rule there. Our ideas will, in the end, shape the world all the better for maturing a little. [*CPW* iii. 282.]

Arnold's view of history as an oscillation or alternation of intellectual principles emerges clearly in this passage as the basis for a 'disinterested' criticism. It incorporates the idea crucial to what may be called his strategy, an idea congruent with certain Christian concepts of the afterlife and with later economic arguments for incomes policy, that a restraint or sacrifice now ensures a repayment at a later date. The importance of the idea is in its assertion of the long-term and circuitous aim of Arnold's cultural project. For it is easy to recognize his short-term policy of disengagement from practical politics and movement into serener realms of sweetness and light without seeing also that this policy is designed, in the long-term, precisely to 'shape the world all the better'. He intends to have an effect upon the world, but an effect that is necessarily delayed by its detour through history. Nor indeed does Arnold really abstain from the choices demanded by the practical world, to the extent that he applies his maxim 'force till right is ready'. His 'Criticism' is to create a current of true and fresh ideas, 'and to leave alone all questions of practical consequences and applications, questions which will never fail to have due prominence given to them' [*CPW* iii. 270]. Here, practical activity is handed over to those who already give it prominence — the existing practitioners, the existing practical authorities. While 'right' matures in quarantine, the established order is the legitimate one.

As part of his withdrawal from the sphere of practice, it is to be noted that the term 'practical criticism' appears in Arnold's writings as a description of the worst kind of criticism, an interested criticism tied to one or another class or political faction in society.

A polemical practical criticism makes men blind even to the ideal imperfection of their practice, makes them willingly assert its ideal perfection, in order the better to secure it against attack; and clearly this is narrowing and baneful for them. If they were reassured on the practical side, speculative considerations of ideal perfection they might be brought to entertain, and their spiritual horizon would thus gradually widen. [*CPW* iii. 271.]

Arnold's reassurance to those who control practical affairs is that he undertakes not to challenge their prerogatives. Yet such reassurance is, unavoidably, a commitment within the practical world, a commitment to defend force, the existing order of things, till right is ready. Under any sustained cross-examination his 'disinterestedness' breaks down. Thus, when he writes, of ideas and practice, that there is good in each and that 'neither is to be suppressed' [*CPW* iii. 265], the pose of neutrality between the two which he adopts conceals the fact that the pact of non-interference in practical affairs does involve a suppression of any new encounter between ideas and practice, thereby reducing ideas to impotence. At one moment Arnold will call for fresh ideas, and at another ask for these ideas to go stale before they can be allowed out into the street. The strong traditions of English anti-intellectualism, of which he makes so much fun his writings, are thereby left unchallenged:'The Englishman has been called a political animal, and he values what is political and practical so much that ideas easily become objects of dislike in his eyes. . . . This would be all very well if the dislike and neglect confined themselves to ideas transported out of their own sphere and meddling rashly with practice . . .' [*CPW* iii. 268), but Arnold is pleading for a place only for ideas-in-themselves.

It has been claimed that Arnold 'wrote the classic protest against Victorian anti-intellectualism'.[1] While many of his catch-phrases about fresh ideas and knowing the best that has been thought in the world may appear to support such a claim, it is still unfounded. The major twentieth-century critics who have most relied upon his work — T. S. Eliot, I. A. Richards, and F. R. Leavis — have recognized his anti-theoretical tendencies as an integral part of his contribution to criticism. Eliot observes, in *The Use of Poetry and the Use of Criticism,* 'In philosophy and theology he was an under-graduate; in religion a Philistine. It is a pleasanter task to define a man's limitations within the field in which he is qualified; for the definition of limitation may be at the same time a precision of the writer's excellences.'[2] He refers here to his condemnation of Arnold's thought, in *The Sacred Wood,* as circular and inconsistent. F. R. Leavis took up the same point in 1938:

The lack of the 'gift for consistency or for definition' turns out to be compensated, at his best, by certain positive virtues: tact and delicacy, a habit of keeping in sensitive touch with the concrete, and an accompanying gift for implicit definition — virtues that prove adequate to the sure and easy management of a sustained argument and are, as we see them in Arnold, essentially those of a literary critic.[3]

Both critics are referring to a curious complicity of strengths and weaknesses in Arnold's writings whereby theoretical vagueness becomes a positive advantage. It seems to these critics that despite his repeated rejection of practical concerns, he was a more 'practical' critic than his stated enthusiasm for fresh ideas and intelligence would imply; practical both in the sense of putting aside logical obstacles to the 'sure and easy management of a sustained argument', and in the sense of intending, ultimately, to change the wider world. The two senses merge when Arnold discusses, in 'The Function of Criticism . . .', the problem of reassuring the practical political people: his only hope of practical influence lies, paradoxically, in the advertisement of his practical neutrality. He complains that the radical writer William Cobbett, like Carlyle and Ruskin after him, has little credibility or effect, 'blackened as he is with the smoke of a lifelong conflict in the field of political practice'. Where, Arnold asks, 'shall we find language innocent enough, how shall we make the spotless purity of our intentions evident . . .?' [*CPW* iii. 275].

It was through this search for an innocent language that Arnold contributed most to the development of a new sense of practical purpose and a new status for English literary criticism. He was to create a new kind of critical discourse which could, by its display of careful extrication from controversy, speak from a privileged standpoint, all other discourses being in some way compromised by partial or partisan considerations. This achievement, gratefully acknowledged by Eliot and Leavis, came at a certain price: a systematic suppression of theory and of argument. This, it will be argued below, is the process of 'compensation' indicated by Leavis, a transaction whose logic can be traced across Arnold's entire prose work.

Arnold's prose writings fall into three overlapping but recognizable phases: the early, more 'classicist', literary-critical writings, the social and theological works of the late 1860s and early 1870s, and the later literary works. The following survey abides by this scheme (including the educational writings with those of the middle period), although, as will be seen, there is no hard and fast distinction to be drawn between periods of a production that shows a remarkable unity of outlook.

ii. *Intellectual deliverance*

Like those of many other critics, Arnold's first public statements on literature flow from his own problems as a practising poet; but he is exceptional in that this commentary upon his own work takes the form not of special pleading or self-advertisement (as it had with Dryden, for example), but of self-rebuke at the shortcomings of his own productions. The first step in a long critical career was taken in 1853 with the publication of a Preface to his *Poems,* in which he attempts to explain why he had not republished 'Empedocles on Etna', a poem which had taken him two years to write. Even before this, several letters to his friend Arthur Clough had spelt out what Arnold considered to be wrong with his own verse, and with modern poetry in general. Against the widely accepted poetic norm of lyric sincerity, of personal utterance, Arnold was feeling his way towards a poetic which could go beyond the morbid agitation of modern verse.[4]

Thomas Carlyle had told his readers to close their Byron and instead open their Goethe, and Arnold seems to have followed this advice, finding in Goethe (as in Sainte-Beuve) a rejection of the Romantic 'sickness' in favour of classical sanity. Arnold diagnoses the sickness as self-absorption and what he calls 'depth-hunting'. In his view, trying to solve the riddle of the universe in poetry is as harmful as merely reproducing life superficially. If poetry is to hold out against the world's multitudinousness, to preserve what he later called the 'inevitability' of verse, the subversive probings of the depth-hunter must be replaced by a new outlook: 'Not deep the poet sees but wide.'[6] Poetry must follow not the vertical axis of probing expressive intensity, but the horizontal axis, by inclusiveness of content, seeing the world steadily and seeing it whole. It is important to note here that Arnold's distaste for analytic 'depth-hunting' was not confined to poetics but seemed applicable to other spheres: 'I for my part think that what Curran said of the constitution of the state holds true of individual moral constitutions: it does not do to lay bare their foundations too constantly.'[7]

The 1853 Preface itself expresses regret at the presence in modern poetry of excessive depth-hunting tendencies, particularly the modern 'dialogue of the mind with itself', whose troubled curiosity of expression Arnold compares unfavourably with the ancient Greeks' calm sense of artistic wholeness. The remedy is for poets to subordinate their expression to an already given content, while a study of the ancient authors is to have 'a steadying and composing effect upon their judgement, not of literary works only, but of men and events in general' [CPW i. 13]. Arnold challenged the simple 'spasmodic' poetics of his time to recognize the demands of what we might now call 'objective form' — the benefits of a 'particular, precise and firm' poetry, in favour of which the poet must try 'effacing himself' [CPW i. 2, 8].

This was one of the first stirrings in modern English criticism of a new 'classicist' strain, which was to flourish openly over sixty years later in the work of T. S. Eliot. In its recognition of the lop-sidedness of the Romantics' vision, and in its attempts to realign a poetic tradition beaten out of shape by the power of these poets, it promises a certain reinvigoration; but only at the cost of an ominous weariness with explanation or argument. Arnold's newly chastened poet 'will not, however, maintain a hostile attitude towards the false pretensions of his age: he will content himself with not being overwhelmed by them. He will esteem himself fortunate if he can succeed in banishing from his mind all feelings of contradiction, and irritation, and impatience.' [CPW i. 14.] This ostrich-like gesture of wounded withdrawal, so familiar in his poetry, indicates a recurrent tendency in Arnold's work, one which becomes more prominent the greater the irritations and contradictions encountered.

One major question raised in the 1853 Preface was the long-standing critical debate over the Ancients versus the Moderns. Arnold felt that the Ancients with their serene and rounded view of the world were still more satisfying than the brilliant but unbalanced Moderns, Shakespeare included. In arguing this case, he was able to draw upon a version of history as a spiral or cyclical movement in which the basic features of epochs widely separated in time could come to resemble each other closely. This model of history, drawn from Vico and Hegel and elaborated by his father Thomas Arnold, among others, allowed Arnold to see earlier civilizations as more modern in many ways than later ones. The Athens of the fifth century BC, Shakespeare's England, and nineteenth-century Europe all faced, in his view, similar problems of transition from heroic or aristocratic to more democratic forms of government; a transition which finds its echo in Arnold's verse as a repeated dirge for past cultures, dead poetic heroes, and (in 'Balder Dead') dead gods.[8]

In his prose work, this historical model appears as the basis for the postponement of creative production until after the 'critical age', and for the maintenance of 'force till right is ready', as well as justifying his preferences for classical authors. It is most clearly expressed in Arnold's inaugural lecture as Professor of Poetry at Oxford in 1857, 'The Modern Element in Literature'. Here he explains the enduring interest of the age of Sophocles by its being 'one of the modern periods in the life of the whole human race' [CPW i. 23], a highly developed and various civilization similar in its complexity to nineteenth-century Europe. Compared with fifth-century Athens, Elizabethan England was coarse, despite its great poetry, while imperial Rome, though undoubtedly a modern civilization, failed to produce a poetry adequate to itself. It is with this problem of 'adequate' poetry, a poetry which can live up to the demands of its age, that Arnold is concerned. More particularly, what the present age demands, he says, is an 'intellectual deliverance':

The demand arises, because our present age has around it a copious and complex present, and behind it a copious and complex past; it arises, because the present age exhibits to the individual man who contemplates it the spectacle of a vast multitude of facts awaiting and inviting his comprehension. . . . [The required intellectual deliverance] is perfect when we have acquired that harmonious acquiescence of mind which we feel in contemplating a grand spectacle that is intelligible to us; when we have lost that impatient irritation of mind which we feel in presence of an immense, moving, confused spectacle which, while it perpetually excites our curiosity, perpetually baffles our comprehension. [*CPW* i. 20.]

The deliverance, Arnold insists, is one of comprehension. Yet doubts about this must immediately be raised by his very terms; for he describes the process so much in terms of Greek tragedy – a grand spectacle in which we acquiesce – that he virtually invites us to confound the end with the means, the task of comprehending our age with the pleasure of witnessing 'this mighty agent of intellectual deliverance', the Greek drama. Might not the sense of harmonious acquiescence be gained more easily by reading Sophocles than by trying to understand all the irritating disputes of modern life? Arnold's thirst for harmony has its own kind of impatience, an impatience with argument which casts doubt upon the nature of his intellectual deliverance.

In attempting to banish all feelings of contradiction, Arnold seems willing to assume any harmonizing idea of the world, but is then unwilling to justify it. This becomes clearer when he moves on to the offensive against literary irritations in his lectures *On Translating Homer* (1861). Basing his concept of Homeric translation on the edifying, character-forming power of the 'grand style', he flatly refuses to define it:

I may discuss what, in the abstract, constitutes the grand style; but that sort of general discussion never much helps our judgement of particular instances. I may say that the presence or absence of the grand style can only be spiritually discerned; and this is true, but to plead this looks like evading the difficulty. My best way is to take eminent specimens of the grand style, and to put them side by side with this of Scott. [*CPW* i. 136.]

So while the mass of quotations may give this work the appearance of a close analysis, it really represents no more than a nakedly comparative *fait accompli*, as can be seen again here: 'Then I ask the scholar, — does Homer's manner ever make upon you . . . an impression in the slightest way resembling, in the remotest degree akin to, the impression made by that passage of the mediaeval poet? I have no fear of the answer.' [*CPW* i. 121.]

While having no reason to fear stage-managed answers, Arnold does harbour a fear of any depth-hunting analysis which might undercut impressionism and force him to disclose and justify the principles of his position. Refusing any disclosure of his reasoning, his argument against what he sees as the artificial 'quaintness' of Francis Newman's translation of the *Iliad* is reduced to a mere point-by-point denial of Newman's views. Convinced that Newman's antiquarian approach cannot render the unity of Homer's poetic effect, Arnold curiously confines his attack to playing off an alternative list of adjectives (plain, noble, rapid) against Newman's descriptions of the Homeric style. In this early use of the empirical 'touchstone' method, he offers a collection of tiny poetic fragments in the name of concreteness ('Is that quaint?'), and so by refusing sustained analysis he reproduces the same impressionistic reliance upon single lines and details in his criticism against which he had always argued in poetry. In place of an open intellectual controversy (because it 'checks the free play of the spirit' [*CPW* i. 169]) Arnold installs an intuitive trade secret: '[Newman] is entirely right when he talks of my ignorance. And yet, perverse as it seems to say so, I sometimes find myself wishing, when dealing with these matters of poetical criticism, that my ignorance were even greater than it is. To handle these matters properly there is needed a poise so perfect that the least overweight in any direction tends to destroy the balance.' [*CPW* i. 174.]

This preference for poise over knowledge contains in germ many ideas extended in Arnold's work of the 1860s. Intuitively assuming his own centrality, he imposes the criterion of 'adequacy' in the negative sense, condemning Newman's erudition for going beyond the harmonious balance which Arnold takes as his starting-point, for being too much of a good thing. In the same way as the violence of the French

Revolution and the fantastic obscurity of the Elizabethans are attributed by him to a premature release from feudal absolutism, so certain kinds of knowledge are worse than balanced ignorance if they threaten Arnold's chosen harmony. 'Terrible learning, which discovers so much!' [*CPW* i. 185.] This was Arnold's verdict on the work of Francis Newman. Even more terrible in his eyes was *The Pentateuch and the Book of Joshua Critically Examined* (Part I, 1863) by William Colenso, Bishop of Natal. This book, the work of a keen mathematical mind, disproved many of the facts and figures given in the Bible by demonstrating their arithmetical impossibility; without, Arnold complained, providing anything to fill the void left by such a demolition. Arnold's essays 'The Bishop and the Philosopher' and 'Dr. Stanley's Lectures on the Jewish Church', attacked the public airing of such well-documented doubts. For those already aware of the Bible's fallibility it was redundant, and for the general public it was positively harmful — 'neither milk for babes nor strong meat for men' [*CPW* iii. 53]. Spinoza, on the other hand, had had the good sense to discuss these questions only in Latin, not in front of the children.

Arnold argues that knowledge and truth are not attainable by the mass of people 'until [a] softening and humanising process is very far advanced' in its operation on them. Until then, new ideas must filter down slowly to avoid convulsion; and if these ideas run ahead of the humanizing process, 'they are even noxious; they retard [its] development, they impair the culture of the world' [*CPW* iii. 44]. The corollary of this is that the guardian of culture will, in turn, seek to retard the development and dissemination of new ideas, keeping them in quarantine until a new harmony can be created:

Some day the religious life will have harmonised all the new thought with itself, will be able to use it freely: but it cannot use it yet. And who has not rejoiced to be able, between the old idea, tenable no longer, which once connected itself with certain religious words, and the new idea, which has not yet connected itself with them, to rest for a while in the healing virtue and beauty of the words themselves? [*CPW* iii. 81.]

This problem of old wine and new bottles is the recurrent theme of Arnold's first series of *Essays in Criticism* (1865).

'What we have to study,' he says of old institutions and ways of thinking, 'is that we may not be acrid dissolvents' of them.[9] The hint given in the debate with Colenso — that anything disruptively premature would need to be restrained — tends to indicate that Arnold was playing the role of a preserver of the old forms of thinking, rather than of a dissolver. It was for *Essays in Criticism* that 'The Function of Criticism at the Present Time' was written, providing with its maxim 'force till right is ready' not only a guide to Arnold's view of social institutions, but an indication of the limits placed upon his proposed intellectual deliverance. The conservative argument of *Essays in Criticism* is closely tied in with Arnold's poetics: both urge us to rely upon the harmony of an already given state of affairs and to regard any further mental activity with suspicion. This has the result of reducing comprehension to the mute contemplation of a spectacle: logicians, Arnold explains, 'imagine truth something to be proved, I as something to be seen; they something to be manufactured, I something to be found'.[10]

Like his critical touchstones, then, Arnold's truths just happen to be *there*. As in the nakedly comparative method of *On Translating Homer,* the free play of the mind really consists in the passive reception of contrasts: placed side by side with the Catholic Church, the English divorce court is hideous — but Arnold noticeably refrains from deciding which attitude to divorce — the Protestant or the Catholic — is actually preferable or right. The question is reduced to one of good taste, as it is again when Arnold's first reaction to the case of a young woman called Wragg accused of murdering her illegitimate child is to deplore the ugliness of her name.[11] Yet the method involved in this is essential to Arnold's project, for the manufacturing of truth involved in adducing the contrasts is easily disguised as the registering of the empirically self-evident, and the 'disinterested' critic is, by this sleight of hand, placed above the modern disease of polemic and partisanship: 'I wish to decide nothing as of my own authority; the great art of criticism is to get oneself out of the way and to let humanity decide.'[12] Facts, those innocent objects upon which we seem merely to stumble, speak their innocent

language apparently of their own accord, while their ventril-
oquist is 'out of the way'.

Arnold told the Royal Academy in 1875: 'My life is not
that of a man of letters but of an Inspector of schools.'[13] He
was both, of course, but it is worth being reminded of the
professional context of Arnold's writings. The inspectorate
which he joined in 1851 was a highly controversial government
department trapped from its very inception in running
battles between Government, Church, and Dissent over the
minds of the rising generations. Problems of the Church, the
state, and education arose simultaneously in his every daily
task, and they overlap accordingly in his writings. The crucial
term 'Culture' itself is often used by Arnold as a translation
of the German *Bildung,* usually rendered as 'education' or
'training'. If Dr Arnold of Rugby tended to see a school as
a microcosm of society, his son often seems to envisage society
as a large school in which the exemplary conduct of teachers
and monitors is decisive: 'It seems as if few stocks could be
trusted to grow up properly without having a priesthood
and an aristocracy to act as their schoolmasters at some time
or other of their national existence.'[14]

Arnold's early work had discussed literature in terms of
models and examples. The terms are the same when he comes
to discuss priests, aristocrats, and schoolmasters. It is the idea
of formative example — similar to his father's emphasis on
character-building — which distinguishes Arnold's views on
education. In his school reports and in the public debates
about state funding of education he deplored the Gradgrindery
of crammed instruction and the Liberals' short-sighted
educational policy of 'payment by results'. In 1862 he risked
his job by directly attacking the *laissez-faire* principles of his
employers' Revised Code (a two-fifths cut in the education
budget which tied capitation grants to examination results),
starkly opposing the demands of social stability to the false
economies of Liberal policy:

Limit the State's duty, in the schools of the nation, to offering a capitation grant for every good reader, writer and cipherer? You might as well limit the State's duty, in the prisons of the nation, to the offering a capitation grant for every reformed criminal! But in prisons, it will be said, the State has another interest besides the reformation of the criminal — the protection of society. We answer: And so, too, in schools the State has another interest besides the encouragement of reading, writing, and arithmetic — *the protection of society*. It has an interest in them so far as they keep children out of the streets, so far as they teach them — the dull as well as the clever — an orderly, decent, and human behaviour, so far as they civilise the neighbourhood where they are placed . . .[15]

Conceiving of education as a civilizing agent rather than just a transfer of information, Arnold defended the idea of formative training, of contact with good literary models in particular, in the hope that a new trained body of teachers could be brought 'into intellectual sympathy with the educated of the upper classes'.[16] He was looking above all for an example to lead the multitude now that priesthoods and aristocracies were losing their power. Given this outlook, it should be no surprise that it is in his report on *The Popular Education of France* (1861) that his ideas on society are most concisely presented. Here he appeals to the middle classes to fulfil their duty as the 'natural educators' of the eager and irrepressible working masses, but he fears that in their present state these middle classes would 'almost certainly fail to mould or assimilate the masses below them' [*CPW* ii. 26], and that the result would be anarchy. The middle classes must embrace state education, not just to assimilate the masses, but to cure their own lack of exemplary governing qualities, otherwise 'a great opportunity is missed of fusing all the upper and middle classes into one powerful whole' [*CPW* ii. 88].

The forging of such a social bloc to meet the challenge of democracy Arnold sees as the task of the state, which he describes (again in the language of character-building) as our 'best self'. This description is important, because in his search for an authoritative example to replace the dying 'feudal habits of subordination'[17] Arnold aims insistently to forge a direct link between state and individual conscience, as here in a letter of 1866: 'Whereas in France, since the Revolution, a man feels that the power which represses him is the *state,* is *himself*, here a man feels that the power which represses him is the Tories, the upper classes, the aristocracy, and so on.'[18] Bringing this individual conscience into immediate relation with the state, and dismissing all that lies between (classes, political rights, independent institutions) as just so much fetishized machinery, Arnold is able to conjure up immediate spiritual causes for civil strife, as in this passage from *Culture and Anarchy* (1869), a work haunted by the fear of large working-class demonstrations of the kind which had damaged the railings in Hyde Park in 1867:

And as to the Populace, who, whether he be Barbarian or Philistine, can look at them without sympathy, when he remembers how often, — every time that we snatch up a vehement opinion in ignorance and passion, every time that we long to crush an adversary by sheer violence, every time that we are envious, every time that we are brutal . . . every time that we trample savagely on the fallen, — he has found in his own bosom the eternal spirit of the Populace. [*CPW* v. 144-5.]

The apparently confessional, self-critical approach here enables Arnold to identify the working class with the very heart of darkness. The far-fetched vocabulary of spiritual forces in *Culture and Anarchy* extends to the state too: it is 'sacred', it is 'one, and noble, and secure, and peaceful', and (getting closer still to Arnold's attenuated definition of God as a 'tendency not ourselves') its proceedings reach 'far beyond any fancies and devisings of ours' [*CPW* v. 223-4. 239].

It might seem odd that the unmediated leap so often made in this work between the heights of state power and the depths of the psyche is passed off so calmly as common sense. Yet Arnold is serious when he claims to be 'a plain, unsystematic writer' keeping 'close to the level ground of common fact' [*CPW* v. 137, 192]. For he is still under fire from the systematic logicians who imagine truth to be something to be proved; and his amateur sociology can stand only by a resolute refusal of systematic reasoning. As Lionel Trilling has shown, Arnold was trying to treat social classes as if they had no distinct interests, when it is precisely their interests and the conflicts between them which make them

classes.[19] This method of 'disinterestedness' gives each class
a tribal nickname (Barbarian, Philistine, Populace) and then
reduces social conflict to a series of imperfections within
each tribe's soul, thus bypassing the question of their social
relations.

Exactly the same method is used in the treatment of
racial characteristics in *On the Study of Celtic Literature*
(1866). Here again, perfection is to be reached by a reshuffling
of permanent racial elements (honesty, charm, intelligence,
etc.) within the national spirit. Every social evil can now be
traced directly to a particular failure of character. So when
discussing the Liberal ideology of free trade in *Culture and
Anarchy,* Arnold seems to penetrate the myths of 'Cheaper
Bread' and expanding populations, but he turns this social
question back into a question of individual self-restraint: 'so
is no man to be excused for having children if his having
them makes him or others lead [an ignoble life]. Plain thoughts
of this kind are surely the spontaneous products of our
consciousness.' [*CPW* v. 219.]

Arnold's influence as a social commentator rests upon his
ability to make these reductions of social to moral tendencies
indeed the spontaneous products of our consciousness, to
make them seem as natural as possible. Such reductions,
dissolving all social problems in a spiritual cloud, reach their
logical conclusion in the later essay on 'Equality' (1878):
'But a community having humane manners is a community of
equals, and in such a community, great social inequalities
have really no meaning, while they are at the same time
a menace and an embarrassment to perfect ease of social
intercourse.' [*CPW* viii. 289.] Taking a united community for
granted, Arnold concludes that any change or movement in
society is a manifestation of the qualities of race and character,
a result of the spiritual oscillation between the racially based
principles of intellectual flexibility ('Hellenism') and moral
strictness ('Hebraism').

In *Culture and Anarchy* Arnold called Christianity 'Hebraism
aiming at self-conquest and rescue from the thrall of vile
affections, not by obedience to the letter of a law, but by
conformity to the image of a self-sacrificing example' [*CPW*
v. 169]. This notion of example as opposed to mechanical

obedience had been developed previously in Arnold's literary, political, and educational writings. Just as the citizen must see the state which represses him as *'himself'*, so the Christian must abandon the idea of fearfully observing the terms of a terrible blood-covenant, and instead imitate Christ's sacrificial method of conscience. In this way 'we are freed from the oppressing sense of eternal order guiltily outraged and sternly retributive: we act in joyful conformity with God's will, instead of in miserable rebellion to it'.[20] Arnold had foreseen the danger of such rebellion against an externalized religion attaching its emotion to the supposed fact, in his attack upon Colenso. He now states clearly the importance of pre-empting the danger: 'It is necessary for the Church, if it is to live, that it should carry the working classes with it.'[21] To this end, the whole vulnerable superstructure (*Aberglaube*) of scripture and theology must be evacuated from the realm of polemical reasoning, and put into the care of literary intuition, as *poetry* with an unassailable truth of its own.

In the urgency of his demand that men should worship together and philosophize alone, Arnold is driven in his theological writings to expel theory altogether, in this logical extension of his maxim 'force till right is ready': 'True, the experience must, for philosophy, have its place in a theory of the system of human nature, when the theory is at last ready and perfect; but the point is, that the experience is ripe and solid, and fit to be used safely, long before the theory. And, it was the *experience* which Jesus always used.'[22] Although it reduces his means of verification to nothing more than the old 'touchstone' method of italicized affirmation, Arnold can in this way insist (in a later essay on Tolstoy) that Christianity is not a set of propositions to be believed, and that Jesus's great merit was that he was 'an opportunist'.[23] In this respect at least, he was devout in his imitation of Christ. With the question of truth again postponed indefinitely, Arnold explains in *Literature and Dogma* (1873) — his most popular work — the advantages of substituting for it the criterion of literary intuition:

The good of letters may be had without skill in arguing, or that formidable logical apparatus, not unlike a guillotine, which Professor Huxley speaks of somewhere as the young man's best companion . . .

But the valuable thing in letters . . . is, as we have often remarked, the judgement which forms itself insensibly in a fair mind along with fresh knowledge. . . . For this judgement comes almost of itself; and what it displaces it displaces easily and naturally, and without any turmoil of controversial reasonings. . . . We are not beaten from our old opinion by logic, we are not driven off our ground; — our ground itself changes with us. [*CPW* vi. 168.]

Arnold saw that the ideology of his day was too brittle and inflexible, too rigidly based upon truth and falsehood, on rules rather than examples, to assimilate successfully the newly enfranchised masses and avoid anarchy. As a remedy, the writings of his middle period demanded, and themselves partly exemplified, a new 'innocent language' immune from 'any turmoil of controversial reasonings'.

Arnold's later essays, which have had a more direct influence upon literary criticism, expound clearly his view of the superiority of poetry to its more abstract rivals. Returning to literary criticism, he brought with him a confirmed distrust of any theory beyond simple empirical contact with the object of study. He dismisses historical and biographical material relating to literature, asserting that Keats's letters to Fanny Brawne should never have been published, and regretting the appearance of Dowden's biography of Shelley, disturbing as it is to his image of 'our original Shelley'.[24] But of all kinds of literary judgement he regards the 'systematic' judgement as 'the most worthless of all':

Its author has not really his eye upon the professed object of his criticism at all, but upon something else which he wants to prove by means of that object. . . . He never fairly looks at it, he is looking at something else. Perhaps if he looked at it straight and full, looked at it simply, he might be able to pass a good judgement on it. As it is, all that he tells us is that he is no genuine critic, but a man with a system, an advocate.[25]

The pretence that critical judgements are self-evident, visible to the naked eye, made in order to attribute sinister motives to other critics, is the fruit of Arnold's developed aversion from political and religious advocacy of all kinds — and one of his lasting contributions to modern criticism. Concluding his introduction to *Isaiah of Jerusalem* (1883), he writes, in the same vein, that 'the right way to get a great author enjoyed is to raise not as much discussion as possible over his meaning, but as little as possible' [*CPW* x. 130]. A similar criterion is applied to poetry itself: conceiving their poetry in their wits rather than in their souls, the poets of the eighteenth century 'composed without their eye on the object'. 'The language of genuine poetry, on the other hand, is the language of one composing with his eye on the object; its evolution is that of a thing which has been plunged in the poet's soul until it comes forth naturally and necessarily.'[26]

Accordingly, Arnold praises Wordsworth's verse as 'inevitable as Nature herself', Chaucer's diction as 'irresistible', Isaiah's prophecies for their 'inexorable fatality', and Byron for letting Nature wield his pen.[27] The argument culminates in the essay on 'Wordsworth' (1879), where much of that poet's work is dismissed as 'abstract verbiage, alien to the very nature of poetry'. He elaborates the distinction in the same essay, looking forward to a time when 'we shall one day learn to make this proposition general, and to say: Poetry is the reality, philosophy the illusion' [*CPW* ix. 49, 48]. Poetry's superiority to philosophy is attributed by Arnold to its 'solidity'. More intellectual than plastic representation, it is more of a 'stay' or support to us than science or philosophy, giving us the idea touched by beauty and (like religion) heightened by emotion. More immediately, it is a firmer support for Arnold than the official dogma of the Church: 'Compare the stability of Shakspeare with the stability of the Thirty-Nine Articles!'[28]

In 'The Study of Poetry', after making his famous claims for poetry's future as a replacement for religion, Arnold asks how we can most benefit from poetry's power to console, interpret, and sustain. For this, we need to recognize the best poetry — discerned not by mere personal preference or according to historical interest, but with a 'real estimate' unencumbered by abstractions: 'Critics give themselves great labour to draw out what in the abstract constitutes the characters of a high quality of poetry. It is much better to have recourse to concrete examples; — to take specimens of poetry of the high, the very highest quality, and to say: The characters of a high quality of poetry are what is expressed *there*.' [*CPW* ix. 170.] Again it is this empirical solidity which makes poetry more reliable than those 'shadows and dreams and false shows of knowledge', religion and philosophy. Even science, for which Arnold showed some enthusiasm in earlier days, will appear incomplete without it — an argument which he goes on to develop in 'Literature and Science' (1882).

In the conclusion to his report on *Schools and Universities on the Continent* (1868) Arnold had weighed the educational claims of science and humane letters in this framework:

The study of letters is the study of the operation of human force, of human freedom and activity; the study of nature is the study of the operation of non-human forces, of human limitation and passivity. The contemplation of human force and activity tends naturally to heighten our own force and activity; the contemplation of human limits and passivity tends rather to check it. Therefore the men who have had the humanistic training have played, and yet play, so prominent a part in human affairs, in spite of their prodigious ignorance of the universe; because their training has powerfully fomented the human force in them. [*CPW* iv. 292.]

Arnold here saw science as if it were not, precisely, a human enterprise. His methods of argument relied upon his pose as a neutral reviewer of the self-evident, always implicitly denying that the acquisition of knowledge was an activity: truth was found not made, consumed but not produced. He is quite consistent, then, when he argues, in 'Literature and Science', that putting science at the centre of education is at variance with human nature on the grounds that after gathering all the facts 'we are still in the sphere of intellect and knowledge' [*CPW* x. 64]. Incapable of guiding our sense of conduct and of beauty, scientific knowledge is 'wearying'. Arnold reaches this conclusion by consistently reducing science to a mere accumulation of facts: 'Following our instinct for intellect and knowledge, we acquire pieces of knowledge,' and so 'one piece of natural knowledge is added to another, and others are added to that, and at last we come to propositions so interesting as Mr. Darwin's . . .' [*CPW* x. 62, 64].

The scientific discipline instils a healthy respect for experience, but its results await the synthetic spark of poetry before they can be consumed in humanized form. That science is much more than a gathering of facts, that it seeks to interpret its own findings in general statements of great human importance, is implicitly denied by Arnold. For him, science is inhuman because of its objects – regardless of its human subject and human motives; and so rendering it human, as he argues in 'On Poetry' (1879), requires handing its results over to another department which can bring the dead facts to life: 'Science thinks, but not emotionally. It adds thought to thought, accumulates the elements of a synthesis which will never be complete until it is touched with beauty and emotion; and when it is touched with these, it has passed out of the sphere of science, it has felt the fashioning hand of the poet.' [*CPW* ix. 62.] The division of labour expounded in 'The Function of Criticism at the Present Time', whereby the philosopher finds the raw materials and the poet synthesizes them, reappears here, no longer as a warning to the poet not to think but as a warning to the scientist not to synthesize facts. Either way the demarcation amounts to a familiar quarantining device against the danger of potentially unsettling new ideas. Science is depicted as an anarchic mass of atomized data, and asked to deliver up any claims to human relevance into the safe-keeping of the poet. Such is Arnold's intellectual deliverance.

From this brief survey of Arnold's prose work, certain general characteristics of his method of intellectual deliverance may be summarized. His radical curtailment of theoretical argument or explanation protects the urgency of empirical demonstration ('*there*') from closer scrutiny, while the evacuated realm of theory thereby becomes vulnerable to reinvasion by cruder substitutes: first a certain vitalism, evident in Arnold's constant appeals to the authority of '*life*' or to 'the instinct of self-preservation in humanity' and its various racially determined manifestations; second, a quantitative calculus which can be found at work in his assertion that morality is 'three-fourths of human life', and in the curious argument that larger nations were more likely to be 'saved' than smaller ones because their enlightened minorities would be larger in absolute numbers.[29] Scientific reason is dissolved in order to install poetry as the only cement of the mind. Critical thought is always in some way inappropriate just at the moment, and is told to wait its turn, as in the astonishing motto Arnold recommended to the students of Liverpool: 'Don't think: try and be patient.'[30]

If Arnold can be said to have combatted anti-intellectualism at all, it was only to make it his own, giving it a public language capable of competing with theory — standing in for it rather than just rejecting it. This is the logic of the compensation referred to by Leavis: consistency of argument is traded in for 'a habit of keeping in sensitive touch with the concrete, and an accompanying gift for implicit definition, — virtues that prove adequate to the sure and easy management of a sustained argument . . .'. Covering the tracks of his arguments beneath supposedly self-evident empirical contrasts while dismissing other positions as premature disturbances, Arnold was able in his prose work to formulate a public discourse, an 'innocent language' which could indeed manage itself with great ease. But such an impregnable position is bought at a heavy price: the 'intellectual deliverance' offered in his inaugural lecture threatens to become a deliverance *from* intellectual — let alone practical — activity, as thought is successively substituted, postponed, concealed, and quarantined.

iii. *The missing centre*

In his attack upon the work of Bishop Colenso in 'The Bishop and the Philosopher' Arnold justified his approach to a book on the Pentateuch with the assertion that theological works had to undergo not only a specialist criticism, but a general one as well.

Literary criticism's most important function is to try books as to the influence which they are calculated to have upon the general culture of single nations or of the world at large. Of this culture literary criticism is the appointed guardian, and on this culture all literary works may be conceived as in some way or other operating. . . . Every one is not a theologian, a historian, or a philosopher, but every one is interested in the advance of the general culture of his nation or of mankind. A criticism therefore which, abandoning a thousand special questions which may be raised about any book, tries it solely in respect of its influence upon this culture, brings it thereby within the sphere of every one's interest. This is why literary criticism has exercised so much power. [*CPW* iii. 41.]

Arnold wishes criticism to have even more power, more authority to appeal to the general public over the heads of particular specialists; to preside over a court or tribunal of intellectual production as a whole. The 'appointed guardian' of general culture is invested with authority to 'deny . . . [a] book the right of existing' [*CPW* iii. 52], while being quite at liberty to ignore the particular substance of the book. As the guardian of culture as a whole, criticism thus stands in a position within culture similar to that of Arnold's ideal state in society: it rises above a multitude of petty particular interests in order to guard against any disturbances to the system as a whole. Within such a framework, Arnold feels justified in suppressing partial viewpoints, in the interests of an assumed totality. Thus when approaching the most pressing problem of practical politics, that of class conflict, he simply expresses impatience with all discussions of capital and labour, since these 'have small existence for a *whole society* that has resolved no longer to live by bread alone'.[31] Characteristically, this banishing of contradictions is achieved in the name of a unity within which economics and politics are only subordinate parts deserving very little attention. So long as this wholeness of society is not accepted, Arnold allows to the state, its highest unifying agent, absolute rights to suppress active dissent. But at the same time he acknowledges the limited moral influence which simple repression has in unifying society, and the danger of the state itself becoming compromised by this course.

It is to meet the need for a wider moral authority as a counterpart to the unifying function of the state that Arnold introduces his concepts of 'Culture', 'Criticism', and 'Best Self'. A supplementary cultural agency is needed throughout society to come to the aid of the state from outside. In *Culture and Anarchy* Arnold identifies this agency as the 'aliens' (elsewhere as the 'remnant') — a minority who distance themselves from their class by pursuing ideals of harmonious perfection:

Natures with this bent emerge in all classes, — among the Barbarians, among the Philistines, among the Populace. And this bent always tends to take them out of their class, and to make their distinguishing characteristic not their Barbarianism or their Philistinism, but their *humanity*. They have, in general, a rough time of it in their lives; but they are sown more abundantly than one might think, they appear where and when one least expects it . . . they hinder the unchecked predominance of that class-life which is the affirmation of our ordinary self, and seasonably disconcert mankind in their worship of machinery. [*CPW* v. 145–6.]

The humanity which characterizes these aliens is a personal microcosm of the unity of society achieved by the state; that is, a rounded, harmonious self-development involving no exaggerated prominence of particular interests or obsessions, particularly the class assertiveness of our ordinary selves. By this standard, religious dissenters, for example, are condemned as 'incomplete and mutilated men' [*CPW* v. 236]. We are all human, Arnold might admit, but some are more human than others.

The problem, which he addressed repeatedly, was that the 'aliens' were so isolated and powerless. How could their influence be spread and class feeling and factional wilfulness further dissolved? One option considered by Arnold was the establishment of an academy along the lines of the Académie Française: an authoritative intellectual centre which could restrain the general intellectual eccentricity of English life. In his essay 'The Literary Influence of Academies' (1864) Arnold argues that the conceited, aggressive ordinary self is given free rein in England because there is no recognized intellectual centre capable of establishing higher standards for people to look up to. The French, on the other hand, have a 'conscience in intellectual matters' [*CPW* iii. 236], a desire to know if they are right to admire a given kind of work, a willingness to defer to a 'central' critical authority. However — and this is a point too often missed by commentators then and since — he finally draws back from advocating the actual establishment of such an academy in England, damaging though he thinks its absence is to the level of cultural life. Instead he proposes that each of us practise a kind of inner restraint: 'But then every one amongst us with any turn for literature will do well to remember to what shortcomings and excesses, which such an academy tends to correct, we are liable; and the more liable, of course, for not having it. He will do well constantly to try himself in respect of these, steadily to widen his culture, severely to check in himself the provincial spirit . . .' [CPW iii. 257].

An understanding of why Arnold 'checked in himself' the strong temptation to establish a real institution of intellectual authority is of particular importance in the history of modern English criticism. Samuel Taylor Coleridge and many later writers had canvassed proposals for an established intellectual 'clerisy',[32] and an academy would appear to be an acceptably modest means for realizing it. Arnold was far-sighted enough, however, to recognize the drawbacks to such a scheme. It was one thing to revere an already established academy sanctified by centuries of tradition, quite another to brave the stormy currents of contemporary English controversy and embark upon the messy practical work of constructing an academy anew. For the benefits which Arnold sees in an academy, one — the French — is enough, as he assures his critics in *Culture and Anarchy*:

. . . it is constantly said that we want to introduce here in England an institution like the French Academy. We have, indeed, expressly declared that we wanted no such thing; but let us notice how it is just our worship of machinery, and of external doing, which leads to this charge being brought; and how the inwardness of culture makes us seize, for watching and cure, the faults to which our want of an Academy inclines us, and yet prevents us from trusting to an arm of flesh, as the Puritans say, — from blindly flying to this outward machinery of an Academy, in order to help ourselves. [CPW v. 234.]

The inward condition of culture, and the inner restraint encouraged by it, are more valuable to Arnold than a new institution which would in all likelihood be overrun by Philistines. It is more to his purpose to foster a habit of intellectual conscience and deference by pointing to England's cultural shortcomings and to the need for a centre, than to create a real centre which might fall short of the ideal. To paraphrase Voltaire's famous remark about God, it could be said of Arnold's intellectual centre that its most important attribute was not its actual existence but an acceptance of the need to invent it.

While there was no need for new institutions to be founded if a sense of their absence could serve the same chastening purpose, the integrity of already established institutions was, for Arnold, essential. Despite his private reservations about the Church of England, he supported it vehemently against the religious dissenters, emphatically asserting that 'The great works by which, not only in literature, art, and science generally, but in religion itself, the human spirit has manifested its approaches to totality and to a full, harmonious perfection, and by which it stimulates and helps forward the world's general perfection, come, not from Nonconformists, but from men who either belong to Establishments or have been trained in them.' [*CPW* v. 237.] Arnold's view of the state as 'sacred' has already been mentioned. He lavished similar extremes of adulation upon Oxford University — a particularly clear example of the lengths to which he went in creating an imaginary spiritual and intellectual 'centre' for English culture. In his private letters and in some of his lesser-known educational reports Arnold admitted that Oxford was little more than a glorified finishing school for the Barbarians, with a stagnant intellectual life.[33] Yet in his more public pronouncements, all his rhetorical gifts are brought into play to weave a myth of a completely different Oxford. This Oxford, despite the passionate theological controversies of Arnold's own student days, is described as 'so unravaged by the fierce intellectual life of our century, so serene!' This serenity, whether or not it corresponds to the real Oxford, is clearly important to his purpose. As he remarks in the Preface to *Essays in Criticism,* 'what example could ever so inspire us to keep down the Philistine in ourselves . . .?' [*CPW* iii. 290]. It is as an edifying example before which our ordinary self is humbled that Arnold values Oxford, as he values the state, the church, and the 'grand style'. The curious lyricism

with which he celebrates Oxford can be accounted for satis-
factorily only by recognizing this category of the exemplary
in his work. The principle of regulating conduct according to
a conscience or Best Self requires an elimination of all external
'machinery' to allow a single exalted model to emerge capable
of attracting the Best Self upwards towards it, out of the
morass of self-will and class feeling. This is the principle
which Arnold applies in religion, where theological dogma is
cut away to reveal the tendency 'not-ourselves' for righteousness exemplified by Jesus's sacrificial method of conscience;
it is the principle followed in politics, where the state becomes
the sole external counterpart and guarantor of the Best Self,
above and beyond classes; in education and literature Arnold's
emphasis is again on the importance of good models. Just as
he can be said to have 'mythologized' Christianity, reducing
to nothing any questions of truth and falsehood, so he is led
to mythologize the state and, in the realm of intellectual life,
Oxford, endowing them with an impossible permanence and
serenity.

The idealization of Oxford which Arnold effects is a
valuable indicator of his own aims — almost an 'objective
correlative' for the long-term project to which he was committed. This becomes particularly clear in his valedictory
lecture in 1867, later incorporated into *Culture and Anarchy*,
where he and Oxford are linked by a more than authorial 'we':

Oxford, the Oxford of the past, has many faults; and she has heavily
paid for them in defeat, in isolation, in want of hold upon the modern
world. Yet we in Oxford, brought up amidst the beauty and sweetness
of that beautiful place, have not failed to seize one truth, — the truth
that beauty and sweetness are essential characters of a complete human
perfection. When I insist on this, I am all in the faith and tradition of
Oxford. I say boldly that this our sentiment for beauty and sweetness,
our sentiment against hideousness and rawness, has been at the bottom
of our attachment to so many beaten causes, of our opposition to so
many triumphant movements. And the sentiment is true, and has never
been wholly defeated, and has shown its power even in its defeat. We
have not won our political battles, we have not carried our main points,
we have not stopped our adversaries' advance, we have not marched
victoriously with the modern world; but we have told silently upon the
the mind of the country, we have prepared currents of feeling which

sap our adversaries' position when it seems gained, we have kept up our own communications with the future. [*CPW* v. 105–6.]

As a major example of this silent influence, Arnold discusses the opposition waged by John Henry Newman and others at Oxford against 'the vulgarity of middle-class liberalism' and 'the hideous and grotesque illusions of middle-class Protestantism'. Such opposition was doomed to failure in the short term, but it animated subterranean currents of feeling which could gather force in the longer term.

Who will estimate how much all these [currents of feeling] contributed to swell the tide of secret dissatisfaction which has mined the ground under the self-confident liberalism of the last thirty years, and has prepared the way for its sudden collapse and supersession? It is in this manner that the sentiment of Oxford for beauty and sweetness conquers, and in this manner long may it contine to conquer! [*CPW* v. 107.]

This Fabian strategy of long-term attrition is clearly Arnold's own course. Veering away from the problems of practical politics (which he leaves in the safe-keeping of the existing authorities), he can find his way more surely to future influence, once history's pendulum swings back again. The attitude which he attributes to Oxford, closely connected with his twin principles of prematurity and postponement, is one of quietly waiting – the same attitude adopted by Arnold's other Oxford creation, the Scholar Gipsy 'waiting for the spark from heaven to fall'.[34] This elusive figure, haunting the outskirts of Oxford for centuries, captures well the ambiguous stance which won Arnold himself a long-term influence: half Oxford academic, half romantic exile, he is not compromised by the fleshly institution (nor it by him), yet orbits around it as a necessary centre of gravity. Arnold's attitude to established institutions of intellectual or political authority preserves this combination of loyalty and wary detachment as a surer means of communicating with the future than any organized school or coterie of his own. Careful not to embody his ideas in any doctrine or school which might be blown away in immediate controversy, he could be confident that they would be resurrected from their suspended animation at the fringes of powerful establishments, when the time was ripe.

iv. *Hellenizing consistently*

In the twentieth century, Arnold's ideas were to be adopted by some as part of a social and cultural crusade, and to reappear after some delay in the world of practical schemes and controversies. But first their development conformed to Arnold's conception of Oxonian historical detour, moving sharply away from immediate practical effect. It is his place as a 'pivot' of this detour which entitles Walter Pater to consideration within a study of criticism's social functions. In his own right Pater is hardly an outstanding advocate of criticism's wider social responsibilities; indeed, when he suspected that the Conclusion to his *Studies in the History of the Renaissance* (1873) might have actual consequences for his readers' behaviour, he moved to suppress it; and apart from this incident his life was one of undisturbed seclusion. Yet as a mediator between Arnold and the critics of the early twentieth century Pater has (like the unaffected 'carriers' of certain congenital diseases) an importance out of proportion to his direct implication in the issues which they explored. At the very least he is useful in bringing to light certain possibilities of Arnold's renewal of criticism which are only faintly discernible under the shiftings and vaguenesses of Arnold's own writings. He elaborated, in Arnold's own lifetime and briefly beyond, aspects of his work which tend to be smothered or contradicted by their author. Where Arnold found it convenient to prevaricate, Pater was more consistent, as T. S. Eliot observed in his essay 'Arnold and Pater' (1930):

. . . it is surely a merit, on the part of Pater, and one which deserves recognition, to have clarified the issues. Matthew Arnold's religion is the more confused, because he conceals, under the smoke of strong and irrational moral prejudice, just the same, or no better, Stoicism and Cyrenaicism of the amateur classical scholar. Arnold Hellenizes and Hebraicizes in turns: it is something to Pater's credit to have Hellenized purely. [*SE*, 441.]

In this sense, Pater's consistency lay in his adherence to the Hellenist ideal of harmonious perfection even after Arnold began to qualify it in a Hebraist direction. After earning a reputation in the late 1860s with *Culture and Anarchy* and the letters collected in *Friendship's Garland* (1871) as an irresponsible opponent of English sobriety, Arnold appears to have reverted to Hebraic rectitude in 1871, under the shock of the Paris Commune and the Prussian victory over France. These events appeared to him to have revealed a major weakness in the moral fibre of Europe's most Hellenist nation; and one of the first signs of this change was his attack upon Pater for daring to rank Victor Hugo alongside Michelangelo, an attack charging Pater with entertaining 'moral sentiments utterly false'.[35] Without abandoning his view of religion and morality as partial 'sides' to human nature, Arnold after 1871 increased their weight to 'three-quarters' of human life, and muted his advocacy of French and Hellenist virtues. The championship of Hellenism fell to Pater, whose first and most controversial book appeared two years after Arnold's reconsideration.

It is not that Pater represents an 'amoral' attitude against Arnold's new-found moralism. Such a view, which students of Pater have repeatedly had to refute, is more a retrospective amalgam of Pater and Wilde than an accurate judgement of the relations between these writers. Indeed, in reviewing Wilde's *The Picture of Dorian Gray,* Pater could argue that in his levity Wilde 'carries on, more perhaps than any other writer, the brilliant critical work of Matthew Arnold',[36] while criticizing Wilde for losing the moral sense and thereby falling short of harmonious perfection. Here Pater uses Arnold's own criteria to chide another Arnoldian. Pater was a moralist, then, but a moralist who attempted to fuse the moral and the aesthetic more insistently than Arnold. For him the moral dimension of literature lay not in a detachable component of high seriousness (a view unconvincingly appended to his essay on 'Style') but in its heightening or fine tuning of sensation. In the detailed and concrete impressions of characters in literature Pater finds a quality of 'sympathy', and on this quality in *Measure for Measure* he writes: 'It is not always that poetry can be the exponent of morality; but it is this aspect of morals which it represents most naturally, for this true justice is dependent on just those finer appreciations which poetry cultivates in us the power of making, those peculiar valuations of action and its effect which poetry actually requires'.[37] By assimilating moral importance into the quality of aesthetic 'fineness', Pater foreshadows the concerns of those Modernist critics for whom the undoctrinal concreteness of literary images is their value, and refined 'sensibility' a moral quality.

By stressing the concreteness of art and poetry against the abstraction of moral systems, Pater was developing a prominent trait of Arnold's criticism: a sense of the more reliable solidity of works of art, and an insistence upon experience as an alternative to theory. And by enhancing this trend Pater's aesthetic empiricism helps to reflect light back upon the Arnoldian framework of criticism. It was observed above that Arnold's thought breaks down at times into a vitalism on the one hand and an oddly Benthamite calculus on the other. In the light of Pater's exaggerated development of this pattern, it is possible to discern more clearly in Arnold's thought a latent positivism — not the strictly conceived Positivism of Auguste Comte himself, whose mannerisms Arnold ridiculed, but of the broader philosophical movement encompassing (in England) John Stuart Mill and George Eliot. Despite Arnold's objections to the radicalism of Comte's English followers and to Comte's arbitrary whims and terminology, the leading English Comtist Frederic Harrison pointed out that Arnold 'was constantly talking Comte without knowing it', and that *Culture and Anarchy*'s conclusion, 'the summing up of the mission of Culture, is entirely and exactly the mission of Positivism'.[38] Particularly in his efforts to 'humanize' religion, Arnold can be seen to have constructed what T. S. Eliot in another context called 'a Comtism from which all the absurdities have been removed' [*SE*, 482], a more 'central' version of Positivism with its Comtist provincialisms and mannerisms cast off.

If Thomas Huxley's unflattering definition of Positivism as Catholicism minus Christianity be accepted, then there is more than an opening for such a philosophy in Arnold's religious writings and in his projected replacement of religion by poetry. Arnold's tendency to jettison Christian theology in favour of its liturgical consolations has been noted above; and it is this element of his thought which Pater develops furthest, not only Hellenizing but ritualizing consistently.

The precedents for such a development are there in Arnold's writings, waiting to be expanded. In an early essay on 'Pagan and Mediaeval Religious Sentiment' (1864) which seems to have influenced Pater, Arnold writes (discreetly avoiding the first person) that the man of imagination and the philosopher 'will always have a weakness for the Catholic Church' [*CPW* iii. 213-14], and in a passage of the Preface to *Culture and Anarchy* suppressed after the first edition he goes further: 'So, again, if we see what is called ritualism making conquests in our Puritan middle class, we may rejoice that portions of this class should have become alive to the aesthetical weakness of their position, even although they have not yet become alive to the intellectual weakness of it.' [*CPW* v. 522.]

Pater takes this concession to ritualism and extends it, particularly in his novel *Marius the Epicurean* (1885), into a celebration of Pagan and early Christian ceremony, looking forward to the 'aesthetic charm of the catholic church'.[39] The other, complementary, side of positivism, its empiricist narrowing of theory to experience, is further elaborated in the turning-point of the same novel, as Marius rejects the savagery of the gladiatorial games without the aid of moral codes: 'His chosen philosophy had said, — Trust the eye: Strive to be right always in regard to the concrete experience: Beware of falsifying your impressions. And its sanction had at least been effective here, in protesting — "This, and this, is what you may not look upon!"'[40] Marius's achievement is the criticism of life by means of touchstones rather than rules. He has succeeded in cultivating or absorbing a moral sense which functions without the need for abstract reflection. In the same way, art, according to Pater, appeals to and fosters the fusion of concrete experience and intellect: in Greek sculpture, meaning and sensuous form coalesce and 'saturate' one another. This becomes a general rule for all art: 'Art, then, is always striving to be independent of the mere intelligence, to become a matter of pure perception . . . form and matter, in their union or identity, present one single effect to the 'imaginative reason,' that complex faculty for which every thought and feeling is twin-born with its sensible analogue or symbol.'[41] What in Arnold was a prominent tendency to purge literature of abstract, reflective, and discursive elements, becomes in Pater's writings a founding principle, to be passed on in this form to Yeats and to Eliot.

More prominent even than their shared positivist tendencies to ritualism and empiricism, is the continuity between Arnold's and Pater's conceptions of Culture as a general and harmonious perfection of human capacities. Arnold insisted in *Culture and Anarchy* that religion, important though it was, could not lay claim to the 'totality' of human affairs, and that it was one 'side' of human nature only. Instead, Culture itself came to stand for this totality, and a rounded development of interests and faculties took the place of righteous conduct. The Dissenters, taking their faith for the whole of human life, were 'incomplete' people; the followers of Culture would seek self-completion, looking to the Renaissance for inspiration, not just to the Reformation. This ideal of harmonious perfection is conspicuously propounded in Pater's *The Renaissance* — a period (broadly conceived) depicted as 'productive in personalities, many-sided, centralised, complete', when artists and philosophers enjoyed 'participation in the best thoughts which that age produced'[42] — precisely the state of affairs sought by Arnold's ideal Criticism. Thanks partly to Pater, that legendary creature 'Renaissance Man' was to cast his shadow over English criticism well into the twentieth century, on humanist and 'anti-humanist' alike. In Pater's work this many-sidedness is taken as the basis for a moral philosophy propounded in both *The Renaissance* and *Marius the Epicurean*: self-cultivation on all sides, as opposed to the sacrifice of one side to the exclusive claims of one interest or principle. The distinction is drawn in *Marius*: 'The ideal of asceticism represents moral effort as essentially a sacrifice, the sacrifice of one part of human nature to another, that it may live the more completely in what survives of it; while the ideal of culture represents it as a harmonious development of all the parts of human nature, in just proportion to each other.'[43]

The same principle is applied in the Conclusion to *The Renaissance*, where Pater insists that systems and theories which require us to 'sacrifice' any part of our experience have no claim upon us. We should govern ourselves according to inclusive rather than exclusive principles. Elsewhere in *The*

Renaissance he makes the same point, that 'the aim of our culture should be to attain not only as intense but as complete a life as possible'.[44] In pursuing this aim, Pater feels, Greek art in particular can help to answer 'the eternal problem of culture — balance, unity with one's self',[45] a problem particularly acute in the modern world. In his criticism this striving for completeness, this inclusive impulse, finds its most extraordinary expression in Pater's famous meditation on the *Mona Lisa*, in which all human history and ten thousand experiences seem to have been concentrated in a single image. Such reflections form an important node of connection between the 'inclusive' aims of Arnold's early classicism and the concern of early twentieth-century criticism with the omnivorous but balanced literary sensibility.

Arnold had maintained that Culture was the seeking of a general and harmonious perfection, and by 'general' he meant social rather than purely individual; yet at the same time he insisted that it was an 'inward' condition independent of institutional machinery. There was a problem in this formula which Pater, again, helped to throw into sharper relief. If Culture is an inward condition, then it can be interpreted as a purely personal quest for self-cultivation: Pater carries this position to its subjective, almost solipsistic limits in *The Renaissance*. Taking Arnold's definition of Criticism's aim, the seeing of the object as in itself it really is, he subverts it by proposing that as a first step one must know one's impression as it really is. And the first step becomes effectively a substitute, all the more easily because Arnold's 'objectivism' had been grounded on little more than subjective assertion to begin with. Pater again appears more consistent and more honest in his development of Arnold's terms. But this logical development is extremely damaging to any claims for 'general' perfection or for Culture's social effects, as the Conclusion to *The Renaissance* makes clear: experience becomes 'dwarfed into the narrow chamber of the individual mind', with each self-enclosed mind 'keeping as a solitary prisoner its own dream of the world'.[46] Culture, or, as Pater sometimes amends it, self-culture seems now radically inert in its inwardness, and the cultured individual completely ineffective outside the closed circle of his or her own sensations: a mere spectator of life, as Oscar Wilde complained of Marius.

To restore the wider demands of community and a semblance of authority to the concept of Culture, both Arnold and Pater have to resort to external institutions; they both choose aesthetically impressive establishments as the most reliable external prop for inward culture. The importance of such institutions for Arnold has already been observed above, but how he ties them, in the Preface to *Culture and Anarchy,* to the ideal of harmonious perfection should also be noticed:

One may say that to be reared a member of a national Church is in itself a lesson of religious moderation, and a help towards culture and harmonious perfection. Instead of battling for his own private forms for expressing the inexpressible and defining the undefinable, a man takes those which have commended themselves most to the religious life of his nation; and while he may be sure that within those forms the religious side of his own nature may find its satisfaction, he has leisure and composure to satisfy other sides of his nature as well. [*CPW* v. 239.]

Arnold's argument appears to be that the fullest development of personality can be achieved most easily by handing over one side of it to an established authority in a form of subcontract. This allows one more leisure to devote oneself to other aspects of self-development. A very similar transaction forms the basis for the hero's tentative adoption of Christianity in *Marius the Epicurean.* In the important chapter 'Second Thoughts', Marius reconsiders the Cyrenaic or Epicurean disdain for religious systems, as an inconsistent rejection of a valuable experience:

But, without him there is a venerable system of sentiment and idea, widely extended in time and place, in a kind of impregnable possession of human life — a system, which, like some other great products of the conjoint efforts of human mind through many generations, is rich in the world's experience; so that in attaching oneself to it, one lets in a great tide of that experience, and makes, as it were with a single step, a great experience of one's own, and with great consequent increase to one's sense of colour, variety, and relief, in the spectacle of men and things.[47]

Marius's return to religion is presented here and increasingly in the later chapters of the novel as an astutely calculated economy. The oddly Utilitarian idea of an economy of sensations as the basis for conduct introduced in the Conclusion to *The Renaissance* is continued in *Marius* in a consistent development towards the adoption of religious observances, if not beliefs. And if we look back to Arnold's explanation of the benefits of established churches, we can see that the argument follows the same course. In both cases dissatisfaction with the isolation of the individual leads the opponent of religion's exclusive claims to revert to orthodoxy, as a convenience. When it is translated beyond the limits of individual self-culture to a social equivalent, the ideal of harmonious perfection comes to rely increasingly upon 'a venerable system of sentiment and idea'.

Taking the achievement of those fleeting 'moments' of aesthetic sensation as the goal of life, Pater offered a more consistent version of the Arnoldian vision of Culture as a replacement for religion. Although he had little to say on criticism's social function, he passed on to twentieth-century critics a number of hints which could be employed in that discussion. The Hegelian concept of the fusion of form and content in art finds its way from Pater to T. S. Eliot's ideas on the unified sensibility, as Pater's vision of the *Mona Lisa* as concentrating all time past and future finds its way to Eliot's concept of timelessness and tradition. To F. R. Leavis he bequeaths his preference for 'such intellectual apprehensions as, in strength and directness and their immediately realised values at the bar of an actual experience, are most like sensations'.[48]

Most of all, Pater foreshadows that other champion of the twentieth-century Arnoldian revival, I. A. Richards. *Principles of Literary Criticism* (1924), in which Richards most clearly spells out his aesthetic and critical doctrines, begins with a strong rejection of 'aestheticism', but it refers not to Pater, only to those who posited the existence of a separate aesthetic faculty cut off from the rest of the mind. Pater was far from proposing anything of the kind; his ideal was precisely the 'synaesthesia', the mental integration to which Richards aspires. His setting of the problem of life as the inclusion of the greatest number of 'pulsations' into the time available is just the problem of Richards's *Principles*.

And when he criticizes the tendency in Wilde 'to lose, or lower, organisation, to become less complex',[49] he speaks precisely the modern language of 'practical' criticism.

NOTES

1. W. E. Houghton, 'Victorian Anti-Intellectualism', *Journal of the History of Ideas*, xiii (1952), 296.

2. *The Use of Poetry and the Use of Criticism* (1933), 105.

3. 'Arnold as Critic', *Scrutiny*, vii (1938), 320.

4. On the prevailing critical orthodoxies, see Alba Warren, *English Poetic Theory 1825-1865* (Princeton 1950).

5. *The Poems of Matthew Arnold* ed. K. Allott (1965), 291 ('Stanzas from the Grande Chartreuse').

6. *The Letters of Matthew Arnold to Arthur Hugh Clough*, ed. H. F. Lowry (1932), 99 (? February 1849).

7. Ibid., 134-5 (1 May 1853).

8. The withering of Arnold's poetry culminated in the production of elegies for dead domestic animals. These are of a certain value, though, in reminding us of his debt to the conservative traditions of German Idealism: his dogs' names were Geist and Kaiser.

9. *CPW* iii. 110 ('Heinrich Heine', 1863).

10. *CPW* iii. 535 (Preface).

11. *CPW* iii. 281, 273 ('The Function of Criticism . . .').

12. *CPW* iii. 227 ('Pagan and Mediaeval Religious Sentiment', 1864).

13. *CPW* viii. 374 ('"Porro Unum Est Necessarium"', 1878).

14. *The Letters of Matthew Arnold*, ed. G. W. E. Russell (1895), i. 115 (20 January 1860).

15. *CPW* ii. 227-8 ('The Twice Revised Code', 1862).

16. *Reports on Elementary Schools by Matthew Arnold*, ed. Francis Sandford (1889), 20.

17. *CPW* v. 118 (*Culture and Anarchy*).

18. *Letters*, ed. Russell, i. 339 (27 July 1866).

19. *Matthew Arnold* (New York, 1939; revd. 1949), 252-3.

20. *CPW* vi. 68 (*St. Paul and Protestantism*, 1870).

21. *CPW* viii. 75 ('The Church of England', 1876).

22. *CPW* vi. 297 (*Literature and Dogma*, 1873).

23. *CPW* xi. 303 ('Count Leo Tolstoi', 1887).

24. *CPW* xi. 326 ('Shelley', 1888).

25. *CPW* viii. 254 ('A French Critic on Goethe', 1878).

26. *CPW* ix. 202 ('Thomas Gray', 1880).

27. *CPW* ix. 52, 174; x. 565.

28. *CPW* ix. 63 ('On Poetry', 1879).

29. *CPW* x. 143-64 ('Numbers', 1884).

30. *CPW* x. 82 ('A Liverpool Address', 1882).

31. *Letters to Clough*, ed. Lowry, 68 (1 March 1848).

32. See Ben Knights, *The Idea of the Clerisy in the Nineteenth Century* (1978).

33. *Letters*, ed. Russell, i. 38, 95; *CPW* iii. 544; *CPW* iv. 318, 323-4.

34. *Poems*, ed. Allot, 338.

35. *CPW* vii. 10-11 ('A French Elijah' 1871).

36. Walter Pater, 'A Novel by Mr. Oscar Wilde', *Selected Writings of Walter Pater*, ed. Harold Bloom (New York, 1974), 263.

37. Walter Pater, *Appreciations, with an Essay on Style* (1889; Library Edition 1910), 184 ('Measure for Measure').

38. Frederic Harrison, *Tennyson, Ruskin, Mill and Other Literary Estimates* (1899), 132.

39. Walter Pater, *Marius the Epicurean: His Sensations and Ideas* (2 vols., 1885; Library Edition 1910), i. 123.

40. Ibid., i. 243.

41. Walter Pater, *The Renaissance: Studies in Art and Poetry* (1873 [as *Studies in the History of the Renaissance*]; Library Edition 1910), 138.

42. Ibid., xiv.

43. *Marius*, i. 121.

44. *The Renaissance*, 188.

45. Ibid., 228.

46. Ibid., 235.

47. *Marius*, ii. 26.

48. Ibid., i. 144.

49. 'A Novel by Mr. Oscar Wilde', 264.

3. A CIVILIZING SUBJECT

'I don't go against our university system: we want a
little disinterested culture to make head against
cotton and capital, especially in the House.'

George Eliot, *Daniel Deronda*

Arnold's decisive contribution to English literary criticism was
a bold extension of its claim to social importance. No less a
task than the prevention of Anarchy now fell to the guardians
of literary culture. But the paradoxical features of Arnold's
Culture, whereby it was most socially useful the more with-
drawn it was from practical affairs, threatened a development
in precisely the opposite direction: the route taken by Pater
away from significant social influence. In particular, Arnold's
early denigration of the 'hideous anarchy'[1] of English literature
in favour of Homer, Sophocles, Dante, Goethe, and a number
of more obscure French writers tended to put literary culture
so far out of the reach of the Philistines (let alone the Populace) that its softening and humanizing effect upon them
would generally be precluded by language barriers.

A specifically English rendering of Arnold's cultural
programme was left, then, to pioneers well outside the
'central' cultural position of Oxford classical studies — in
large measure to the Philistines themselves, and to their
educators. Arnold, indeed, was by no means the first to
draw attention to the possibilities of literature as an antidote
to the narrow class-life of the Philistine. One of the first
school anthologies of English verse and prose, Knox's *Elegant
Extracts*, justified its existence in these terms: 'there is no
good reason to be given why the mercantile classes, at least of
the higher order, should not amuse their leisure with any
pleasures of polite literature. Nothing perhaps contributes
more to liberalize their minds and prevent that narrowness
which is too often the consequence of a life attached, from
the earliest age, to the pursuits of lucre.'[2] Arnold's sense of
literature's uses as an agent of social enlightenment is here

already, suggested by the problems of compiling and justifying the classroom textbook. By the time that Arnold himself was about to enter the discussion, its terms had become clearly recognizable. H. G. Robinson, a teacher at York Training College, wrote in 1860: 'It is, however, in connexion with what is called "middle class" education that the claims of English literature may be most effectively urged. In that literature, properly handled, we have a most valuable agency for the moral and intellectual culture of the professional and commercial classes.'[3] Robinson goes further than merely picking out the 'target' for literary culture; he provides a rudimentary account of its moralizing and liberalizing effects:

The student will learn to appreciate the temper with which great minds approach the consideration of great questions; he will discover that truth is many-sided, that it is not identical or merely co-extensive with individual opinion, and that the world is a good deal wider than his own sect, or party, or class. And such a lesson the middle classes of this country need. They are generally *honest* in their opinions, but in too many cases they are *narrow*.[4]

Arnold was not alone, then, in promoting a particular social mission for literary studies in the battle against the sectarian entrenchment of the middle classes. It was more that his particular location in the educational system (son of Dr Thomas Arnold, brother-in-law of Gladstone's Education Minister William Forster, slum-school inspector, and Oxford Professor, all at once) forced the issues upon him in concentrated form, calling forth a more comprehensive and sustained exposition of the function of criticism and its implications for philosophy, science, religion, and politics. His achievement here was to be a kind of prophecy, a reference point for all future combatants in debates over the uses of literary study — in particular those debates following the First World War, which will be discussed in later chapters. Meanwhile others were at work bringing about that integration of literary culture into institutional forms which laid the basis of twentieth-century English criticism.

The record of literary criticism's incorporation into the workings of higher education, with its own professors, examinations, textbooks, and sub-specialisms, is often

confused either by wistful complaints about 'professionalization' and the disappearance of the amateur, or by a
clutter of amended university statutes, personalities, and
anecdotes. Thus E. M. W. Tillyard in his memoir of English
at Cambridge announces that he is engaged in 'the tricky
business of blending fact with gossip and personalities'[5.] —
a business more likely than not to obscure a more than
superficial understanding of this history and its logic. Important as some of the personalities are, they owed their success
ultimately to the pressure of larger social and educational
developments; among these we can isolate three principal
factors which ensured literary study, in particular of English
literature, a permanent place in higher education. These are
first, the specific needs of the British empire expressed in the
regulations for admission to the India Civil Service; second, the
various movements for adult education including Mechanics
Institutes, Working Men's Colleges, and extension lecturing;
third, within this general movement, the specific provisions
made for women's education. The predominance of these
factors was highlighted in the fierce debate at Oxford over
the proposals for a school of English. The conservative
faction published in the *Oxford Magazine* in 1887 an attack
on the English school's champion, John Churton Collins, in
the form of a drama in which an extension lecturer named
Mr Random Tearem announces: 'The eyes of civilised Europe
are fixed upon me, and to a lesser extent upon the Heb-
domadal Council. A literary education is imperatively demanded for the apostles and exponents of literature in Girton
and the Colonies, in the Provinces and at Newnham, in our
public schools and at Lady Margaret Hall.'[6] Parody here, as it
often does, isolated the principal factors behind the movement, more accurately than Collins himself ever did. It remains,
though, to trace the workings of these highlighted factors,
and to show how they moulded the theory and practice of
English teaching.

The initial openings for English studies in secondary and
higher education came as educationalists realized that sooner
or later greater provision would have to be made for scientific
and technical instruction, given the increasing complexity
of industry and the need for British capital to compete on

favourable terms with Germany and the United States, where technical training was further advanced. The relation of literature to science became an immediate problem for the curriculum. A movement towards modern or liberal studies tended to follow the growth of technical instruction like a shadow, providing the 'human face' to industrial growth in the form of a socially unifying national heritage, both historical and literary.[7] In the Mechanics Institutes in particular, English Literature and technical subjects developed alongside each other.

Many educationalists agreed with Arnold that in all sectors of education the provision of practical knowledge had to be supplemented by a humane, moralizing subject which could harmonize an otherwise anarchic profusion of 'dry facts'. English Literature was an ideal candidate for this position, since it could be offered as a substitute for Classics, acting as a liberal counterweight to technical learning without taking up so much time in basic linguistic drilling or in teaching resources. An anonymous handbook for teachers, *How to Teach English Literature* (1880), after making the familiar assertion that the subject helps to 'promote sympathy and fellow feeling among all classes', adds a more tangible recommendation in its favour: 'But "Literature" has this advantage over some of the so-called 'specific' subjects, that the heaviest part of the work of preparation falls upon the scholar, and makes comparatively small demands upon the teacher's time'.[8] In the influential collection of *Essays on a Liberal Education* (1867), J. W. Hales put forward this conception of the 'poor man's classics': 'In schools whose pupils are not destined to proceed from there to a University, or to a life of studious leisure and opportunity, English should, I think, be made the prominent linguistic and literary study.'[9] In the same volume Henry Sidgwick recommended the dropping of Greek in favour of English and Natural Science; and Robinson had urged the benefits of English Literature 'especially in those schools where it is impossible that the majority of the pupils can ever become good Greek or Latin scholars'.[10] The Newcastle Commission on popular education had suggested in 1861 that pupil-teachers should study English in the same way that public-school boys studied

Latin and Greek. Likewise the Taunton Commission (Schools Inquiry Commission) reported in 1868 a big decline in the provision of Classical teaching in the grammar schools, and again recommended English as a stop-gap. These developments led eventually to passive support from Classicists in the universities, who recognized that schools of English might benefit them by drawing off their weaker students.

i. *The clown in the boudoir*

Arnold had often spoken of the need for a softening and humanizing influence to be exerted upon the masses in Britain, to wean them from class-conflict and intellectual turmoil, and had offered poetry as a means to that end. The introduction of literary study among the middle and working classes through Mechanics Institutes, Working Men's Colleges, and extension lecturing was already taking place with this aim in mind. The first history of F. D. Maurice's Working Men's College, to take just one well-documented example, refers again and again to this primary social objective; Lord Avebury congratulated the college for providing 'one of the good influences which in our country so happily link different classes together. This indeed was one of the main ideas which animated and inspired the founders and early supporters of the College.' Maurice's own hope was that through the college's work 'class may be united to class, not by necessity only, but by generous duties and common sympathies'.[11] The college was to give effect to this fundamental aim of 'Christian Socialism', a doctrine designed specifically to head off the Chartist movement in 1848.

Arnold's view of the importance of popular education for 'the protection of society' has been remarked upon above; he compared the necessity of schools with that of prisons as guarantors of social stability. T. B. Macaulay had followed the same line of thought when he argued that teachers should be trained and paid by the state on the same principle as soldiers.[12] Thomas Hughes, the author of *Tom Brown's Schooldays* and a friend of Matthew Arnold in their student days, put this principle into effect by forming his pupils at the Working Men's College into a patriotic militia and teaching them basic military drill.[13]

Though the need to exercise social control over the urban masses did lie behind the fostering of many adult education projects, this need could by no means be fulfilled by simple regimentation alone. Nor, on the other hand, was adult education designed simply to impart freely to the workers the benefits of a liberal education as enjoyed by their rulers. As one of the latter — a certain Lord Playfair — told the London Society for the Extension of University Teaching in 1894: 'The main purpose is not to educate the masses, but to permeate them with the desire for intellectual improvement, and to show them methods by which they can attain this desire. Every man who acquires a taste for learning and is imbued with the desire to acquire more of it, becomes more valuable as a citizen, because he is more intelligent and perceptive.'[14] This is a fine distinction, but an important one, which becomes clearer the more acquainted we become with the literature of adult education. The aim seems to have been not to dole out knowledge or hand over educational facilities, but to set in motion within the hearts and minds of working people a particular process of self-improvement, or — what was just as beneficial — a feeling of their need to be improved according to others' standards.

F. D. Maurice, Professor of English Literature and History at King's College London, in the 1840s, one of the great instigators of the adult education movement and founder of the model Working Men's College, held as his cardinal principle that every one was a spiritual being 'whether he knew it or not'[15] — the implication being that most people did not know it and would initially have to have their spirits tended for them. Literature played a central part in this campaign for self-cultivation. In the first place it could perform the task of elementary 'political culture', as defined by J. C. Collins, himself an extension lecturer, in his manifesto *The Study of English Literature* (1891): '[The people] need political culture, instruction, that is to say, in what pertains to their relation to the State, to their duties as citizens; and they need also to be impressed sentimentally by having the presentation in legend and history of heroic and patriotic example brought vividly and attractively before them.'[16] Part of this 'political culture' was already provided in adult education, not just by militarists like Hughes, but by the many courses in Political Economy which demonstrated the necessity of collaborative partnership between capital and labour. Patriotic poetry of the kind produced at this time by Henry Newbolt (himself a later champion of English) could give a more easily digestible emotional guarantee to this political education.

Such an application of literary study can of course be seen as relatively simple indoctrination. However, there was a further use for literature, for which the term 'indoctrination' as it is usually understood cannot apply, for it relied upon the process alluded to by Lord Playfair above: the fostering of a more or less independent process of reflection by the learner, rather than the passive swallowing of dogma. H. G. Robinson, whose hopes that literature could overcome the narrowness of the middle class have been quoted above, thought that the dangerous and extravagant imaginations of young people could be cured by literary study as a kind of 'homeopathic treatment'[17] preferable to surgery. The idea is close to Arnold's distinction between drilling or cramming and formative example in education. The Revd Canon Browne in a lecture to university extension teachers attributed the orderliness of their students, in a similar way, to the civilizing effect of study itself, rather than to the actual information imparted:

And, apart from the subject of study, the fact of study should ensure for us the sympathetic help of men who are — to whatever party in the State they incline — worthily in a position of authority and power. There are nations in Europe where the students are a body liable to dangerous explosions of political feeling. . . . But the students whom the statesman sees, when he glances our way, are men and women who spend the little leisure they have from their work in life in quiet, peaceful study; students whose studies predispose them to orderliness . . . students who to the warning lessons of history have added the civilising, softening charms of the noblest literature in the world.[18]

Just how literature in particular has this softening effect, predisposing its students to order, is suggested by Robinson, in a closer analysis of what he had called a homoeopathic treatment:

Assuredly then among the liberal arts that so humanise, standard literature occupies the first place. If anything will take the coarseness and vulgarity out of a soul, it must be refined images and elevated sentiments. As a clown will instinctively tread lightly and feel ashamed of his hob-nailed shoes in a lady's boudoir, so a vulgar mind may, by converse with minds of high culture, be brought to see and deplore the contrast between itself and them, and to make an earnest effort to put off its vulgarity.[19]

Although Robinson would not have known it, his theory had already been confirmed by the experience of the Chartist agitator Thomas Cooper, whose memoirs testify to the sobering effects of literary self-education upon the working-class reader:

All this practice seemed to destroy the desire of composing poetry of my own. Milton's verse seemed to overawe me, as I committed it to memory, and repeated it daily; and the perfection of his music, as well as the gigantic stature of his intellect, were fully perceived by my mind. The wondrous knowledge of the heart unfolded by Shakespeare, made me shrink into insignificance; while the sweetness, the marvellous power of expression and grandeur of his poetry seemed to transport me, at times, out of the vulgar world of circumstances in which I lived bodily.[20]

It was in this way, ideally, that the humanizing power of literature operated to counteract anarchy. If the savage Chartist could be soothed so easily, what wonders could literature not perform in the world? Nearly every theorist of popular literary education in this period attempts to show that great literature is capable of breaking down class differences and showing how unimportant they are. It is Robinson, again, who puts this argument in its full-blown form:

Large views help to develop large sympathies; and by converse with the thoughts and utterances of those who are intellectual leaders of the race, our heart comes to beat in accord with the feeling of universal humanity. We discover that no differences of class, or party, or creed, can destroy the power of genius to charm and to instruct, and that above the smoke and stir, the din and turmoil of man's lower life of care and business and debate, there is a serene and luminous region of truth where all may meet and expatiate in common.[21]

Yet it is on the very same page that Robinson explains the effect of literature on the lower orders in terms of the clown shamed by the boudoir. Far from bringing all classes into the luminous realm of universal humanity, literary education as Robinson sees it actually forces an awareness of class inferiority upon its unrefined readers, making them ashamed of their 'insignificance' before the intellectual leaders of the race, and numbing their own creative capacity. Robinson's serene region of truth is very similar to Arnold's, for it places itself 'above the smoke and stir, the din and turmoil of man's lower life of care and business and debate' and above class and party precisely by suppressing these in the hob-nailed literary student's mind, through a particular form of cultural intimidation. In this way great poetry seems to immobilize its consumers in a contemplative attitude disengaged from their own action and experience. The homeopathic effect of this is well described by another extension lecturer, John Morley, a leading figure in several Liberal governments, writing here of the potential value of literary study: 'Yes, let us read to weigh and consider. In the times before us that promise or threaten deep political, economical, and social controversy, what we need to do is to induce our people to weigh and consider. We want them to cultivate energy without impatience, activity without restlessness, inflexibility without ill-humour.'[22]

ii. *An additional accomplishment*

After extension lecturing, the second butt of the *Oxford Magazine*'s ridicule in its satire on English studies was women's education; for within the general movement for adult education the provision of extension classes and eventually colleges for women was an important influence towards the growth of English studies in particular. Exclusion from scientific training and from the professions for which Latin and Greek were required left women almost entirely restricted to English history and literature or modern languages if they were to embark upon any programme of learning. The records of the Association for Promoting the Education of Women in Oxford, for example, show that classes in these subjects were much better attended than any others. The Taunton Commission reported in 1868 that 'English literature occupies a more prominent position in the education of girls than of boys.'[23] Rather than attempt to reverse the unequal balance of the girls' school curriculum (though a few women did take up scientific subjects through extension schemes), the Ladies Colleges and extension lectures for women tended to reproduce on a grander scale the predominance of French conversation and poetry-reading.

Many of the movement's promoters saw their job in fact as a 'homeopathic' attempt to forestall any more profound change in women's traditional position. F. D. Maurice, who in order to raise the standard of governesses founded the Queen's College for Women in 1848, was quite firm on this question:

In America some are maintaining that they [women] should take degrees and practise as physicians. I not only do not see my way to such a result; I not only should not wish that any college I was concerned in should be leading to it; but I should think there could be no better reason for founding a college than to remove the slightest craving for such a state of things, by giving a more healthful direction to the minds which might entertain it.[24]

These classes for women were not designed to emancipate, but to confirm women in their established roles. 'They need education, not only to show them what they can do, but what they cannot do and should not attempt.'[25] Within this conception of women's education, the 'healthful' value of literary study was that it did not seem to involve straying too far from the acceptable staple of artistic 'accomplishments' which made up the wealthier woman's training for the marriage market. As the Taunton Commission observed, in referring to the education of young women: 'Of all the "solid" subjects . . . those comprised under the name of modern languages and literature are most prized by parents, and ranked next to the accomplishments.'[26] As an additional accomplishment, literary appreciation (preferably with bowdlerized texts)[27] was seen as useful in helping to train a wife in her duties of sympathy and understanding.

Charles Kingsley, in his first lecture as Professor of English at Maurice's Queen's College, expressed the hope that through literature 'woman should be initiated into the thoughts and feelings of her countrymen in every age . . . that knowing the hearts of many, she may in after life be able to comfort the hearts of all'.[28] For Kingsley, English literature provided the finest intimate history of the nation's enduring spirit.

Such a course of history would quicken women's inborn *personal interest* in the actors of this life-drama, and be quickened by it in return, as indeed it ought: for it is thus that God intended woman to look instinctively at the world. Would to God that she would teach us men to look at it thus likewise. Would to God that she would in these days claim and fulfil to the uttermost her vocation as the priestess of charity![29]

Kingsley was expressing here a prevalent opinion that the process of softening and humanizing the middle classes through literary culture would be led by women. This view is echoed in the Taunton Commission's report: 'Mr. Bryce, too, has dwelt on the greater amount of leisure possessed by the women in a mercantile community, if, indeed, it should not rather be said, that it is possessed by them alone; and remarked that we must, therefore, look to them for the maintenance of a higher and more cultivated tone in society.'[30]

Kingsley, and probably Bryce, seem to have had in mind a rather vague spiritual influence, but if we are referring to the final securing of a permanent national provision for the teaching of English Literature — the setting up of schools of English at Oxford and Cambridge — then it is certain that the movement for women's education was a vital factor. More middle-class in composition than the tradition of the Mechanics Institutes, it was able gradually to establish footholds in the universities themselves. In the precarious first five years of the English school at Oxford, it was women who provided the vast majority of candidates for the new examinations (sixty-nine, against only eighteen men), while they made up two-thirds of the Modern Languages school at Cambridge from which the English school there grew up. Again, during the first years of this school, large numbers of male students were away on war service, and although Professor Quiller-Couch always began his lectures with 'Gentlemen . . .', his audience was composed largely of women.

iii. *The native culture*

While many of the pressures for the institutionalized study of English literature came 'from below', seeping up from Mechanics Institutes and the newer colleges before conquering the older seats of learning, there had been an important initiative 'from above', in the recommendations of the Civil Service of the East India Company report of 1855, which outlined plans under the 1853 India Act to open the most lucrative and prestigious administrative posts in the empire to competitive examination. In drawing up their list of examination papers to be set, the committee appointed to report on this wrote: 'Foremost among these subjects we place our own language and literature', and they accordingly allotted a possible 1,000 marks (equalled only by mathematics) to the section on English literature and history, to enable candidates 'to show the extent of their knowledge of our poets, wits and philosophers'.[31] The committee also felt that in imparting 'a taste for pleasures not sensual', literature would help the young administrators to resist the dangers of corruption and 'scandalous immorality' to which their power might expose them.[32]

Britain's overseas interests had stimulated her higher education requirements before: the first provision for modern languages at Oxford had been made at the command of George I in 1724, expressly to turn out twenty fluent linguists for the diplomatic service. The effects of the 1853 India Act did not hit the universities so directly at first, but they were an important precedent, officially encouraging the study of English literature for the good of the empire. Thomas Babington Macaulay, the guiding hand behind the report's recommendations, had long expounded the case underlying such a literary policy. Speaking on 'The Government of India' in 1833, following his appointment to the Supreme Council of India, he had put this point to the House of Commons:

Consider too, Sir, how rapidly the public mind of India is advancing, how much attention is already paid by the higher classes of the natives to those intellectual pursuits on the cultivation of which the superiority of the European race principally depends. Surely, in such circumstances, from motives of selfish policy, if from no higher motive, we ought to fill the magistracies of our Eastern Empire with men who may do honour to our country, with men who may represent the best part of the English nation.[33]

Or, as Macaulay rephrased it for his slower listeners, 'To trade with civilised men is infinitely more profitable than to govern savages.'[34] He looked forward, therefore, to the conscious propagation of 'that literature before the light of which impious and cruel superstitions are fast taking flight on the banks of the Ganges. . . . And, wherever British literature spreads, may it be attended by British virtue and British freedom!'[35]

Members of the India Civil Service contributing to a symposium on 'The Duties of the Universities towards our Indian Empire' in 1884 were to agree on the benefits of such a cultural mission, affirming that 'the culture that men got at Oxford or Cambridge was of the greatest importance in dealing with the natives'.[36] The report of 1855 had decided that knowledge of the languages, customs, and economy of the natives themselves was not to be required of their prospective rulers, since this could be learned by the successful candidates while waiting to catch the boat. It was felt necessary to bear in mind the many unsuccessful candidates who would stay in Britain and would not have lost anything by a study of their own native traditions. The domestic repercussions of the new system were accurately foreseen in the report: 'We think that we can hardly be mistaken in believing that the introduction of that system will be an event scarcely less important to this country than to India. . . . We are inclined to think that the examinations for situations in the Civil Service of the East India Company will produce an effect which will be felt in every seat of learning throughout the realm.'[37] Indeed, this initiative triggered off a spate of public examinations, many of them in English Literature, which acted as a strong lever against the conservatism of the universities' curricula. Oxford began to feel the effects when dons complained that their best students were being distracted from their proper studies by these examinations. The champion of English Literature at Oxford, John Churton Collins, made his living for some time by coaching candidates in English for the Civil Service, and men of letters including Matthew Arnold and Walter Raleigh were proud to act as examiners.

Although it hastened the introduction of formal literary study in English into higher education, the institution of Civil Service examinations in English had a doubtful effect upon the actual methods of teaching literature, because the examinations had been drawn up and set well before any organized teaching, let alone an accepted pedagogic method, could be evolved. English Studies were subordinated to examinations before anyone could really say that English Studies existed. And the deadening form of this subordination can be guessed from the evidence given to the Taunton Commission by one G. W. Dasent: *'With regard to literature, what is exactly the meaning of teaching English Literature?* I understand by teaching English Literature, the reading and remembering as much as you can of as many authors as you can.'[38] Dasent was an examiner for both the India Civil Service and the Council of Military Education.

The three factors behind the movement for English Studies — extension teaching, the colonies, and women's colleges — highlighted by the *Oxford Magazine* were seen as such easy targets for ridicule because they were so clearly 'provincial' (in Arnold's sense), that is, well outside the 'central' tradition of classical university education. Arnold's conceptions of the humanizing and socially healing power of literary culture had in fact quickly taken root where Homer was unavailable: among women, artisans, Indians, and their respective teachers. Having gone among the Philistines,[39] the movement for English literary culture now returned to claim its place at Oxford in a form which invited suspicion and reserve.

The contradictions of this process were concentrated in the person of John Churton Collins, the most vocal campaigner for an English school at Oxford. Collins was very much an orthodox Hellenist influenced by Jowett and the Arnolds, who insisted upon the unbreakable connection between Classical and English literature; his proposals were for an English school open only to trained Classicists. Yet his ungentlemanly Philistine zeal in propagandizing for such a school, combined with his status as a mere extension lecturer and examination coach (quite apart from his personal eccentricities) meant that the Classical dons kept their distance.

Collins's greatest battles, though, were with another group within the university: the philologists and Saxonists, many of whom supported the idea of an English school based on their own studies. Collins would only support a school which was non-philological and tied to the Classics. The debate thus turned into a confused three-cornered fight with many overlapping positions. When the Merton Professorship of English Language and Literature was established in 1884 Collins thought that at last some provision was being made for literature, to complement the already existing Chair of Anglo-Saxon, and he applied for the post. It was awarded, though, to another German-trained philologist, A. S. Napier, whereupon Collins denounced the appointment as a fraud against literature, and embarked upon a series of attacks on philology as a mere 'fact-grubbing' subject. He hardened Arnold's distinction between humanized literary culture and the dead facts of science into a choice between body and soul: 'Philological criticism is to criticism, in the proper sense of the term, what anatomy is to psychology. . . . The scalpel, which lays bare every nerve and artery in the mechanism of the body, reveals nothing further.'[40] While literary criticism brought us into contact with the 'aristocrats of our race', philology mixed with writers of the lowest intelligence. Philology, Collins argued, should loosen its stranglehold upon literature and take its place among the sciences instead.

At this, the philologists took offence, drawing around them many others who were worried by Collins's apparent disregard for scholarship. This group returned caricature for caricature. E. A. Freeman, the Regius Professor of History, predicted that the study of English literature would degenerate into mere 'chatter about Shelley' without philology to act as a corrective discipline, since all literary judgements were a matter of personal taste: 'we do not want, we will not say frivolous subjects, but subjects which are merely light, elegant, interesting. As subjects for examination, we must have subjects in which it is possible to examine.'[41] To complicate the debate further, the ethnological dimension was raked up by the Professor of Moral Philosophy, Thomas Case, as an extreme Hellenist argument against any English school at all: 'An English School will grow up, nourishing our language not from the humanity of the Greeks and Romans, but from the savagery of the Goths and Anglo-Saxons. We are about to reverse the Renaissance.'[42] Others argued that the new school might be of use in getting rid of the weaker Classics students, and that the demands of 'lower' education should be met on this basis, because 'the women should be considered, and the second and third rate men who were to become school-masters'.[43] Plans for a new school of English Language and Literature were approved by Congregation in 1893, with no clear agreement having been reached on its nature or purpose.

In their way, Freeman and the philologists were justified in pointing to the lack of rigorous discipline in purely literary study as it was then conceived. Although for their part they appeared to be defending scholarly discipline largely for its own sake, their anxieties indicated a real shortcoming in the traditions of English study as it had developed outside the universities. It was not that Collins himself was unconcerned about scholarly accuracy — far from it — but that the different educational purposes to which English studies had been moulded left the question of their precise benefits very much taken for granted. Lecturers in English Literature had been content to explain the value of their subject within Arnold's terms of formative example or 'contact' with great minds — a very simple theory of cultural contagion which can be found in most apologies for popular literary education in this period. This conception of literature's value was sufficient for those who wanted to give the workers a chastening example to look up to, but it was to impede the development of any thorough or systematic study of literary works, by concentrating all attention on purple passages or Great Authors considered as higher personalities.

Such 'personalization' was particularly evident in the presentation of literature for women (for reasons touched upon by Kingsley above). The Taunton Commission noted the frequent recitation of famous passages in girls' schools, but remarked also that 'the critical study of a great work in its entirety is not attempted'.[44] F. D. Maurice based his theory of literature's value upon its ability to provide us with personal friends rather than 'mere words'.[45] The form of the extension lecture itself (since it was rarely possible to back this up with close reading of a series of agreed texts) meant that there was a tendency for the lecturer merely to expound the peculiar beauties of one author or another in a more or less biographical manner. Collins at least was well aware, and very critical, of the tendencies towards mere gossip in literary studies. In fact his reason for insisting upon the establishment of English schools at the universities was precisely that they would be able to ensure a firmer organization of the subject and turn it into a discipline proper. But the longer the universities kept their distance, the more this practical organization fell into the hands of other authorities. The Civil Service Commissioners in particular contented themselves with quantifying by examination the extent of their candidates' acquaintance with the nation's great poets, thereby superimposing upon literary studies a form of organization which disrupted them without attempting any conscious reorganization.

When Freeman raised the problem of how to examine in taste, how to avoid simple cramming on the one hand or 'chatter about Shelley' on the other, he was certainly twisting the terms of the debate, but his questions were still not to be answered for decades to come. When the study of English literature got under way in the universities, nobody had delved into these issues in a way which could establish a method of study more satisfactory than that already dictated by the needs of extension classes and the Council of Military Education. Until the work of I. A. Richards in the late 1920s provided an acceptable answer, early teachers of English literature at universities had to sustain what was for many of their critics a barely legitimate subject on makeshift analogies with Classics, or on enthusiasm alone.

iv. *Confessions of a pimp*

After the initial fervour of the movement for the teaching of English literature came a realization that admission to the universities had increased rather than solved the problems. The resulting disillusionment made its deepest impression upon the career of Walter Raleigh, the first really 'literary' Professor of English at Oxford. Raleigh's is a case worth examining closely. He published little literary criticism, so paralysed was he by doubts about his subject (indeed, he once claimed that his book of patriotic essays *England and the War* was worth more than all his literary work), but hidden reflections on literature and criticism can be traced in his collected letters. Here he can be found explaining how he became involved in a profession whose purpose he never fathomed: 'I am a teacher by accident. I had not money enough for the Bar, or anything else, after graduating — second class. I had read a good deal of English literature and philosophy while I was supposed to be reading history, so I got a chance early in the movement for teaching English.'[46]

Raleigh's first job was implementing Macaulay's cultural crusade in India, as Professor of English Literature at the Anglo-Oriental College in Aligarh. He soon found that his task was simply 'to cram a well worn subject into a given number of unfilled heads',[47] and he lost interest in the mechanical system of education which he dubbed the 'Calcutta Mill'. Returning to England, he took up another post at Owens College, Manchester, lecturing in History and Literature, which he found just as cramping and depressing: 'I made some remarks on Poetry in general which cost me more than fifteen matter of fact lectures, and they just laid down their pens and smiled from an infinite height. So I must boil down text books in the recognized fashion.'[48] Frustrated by the limits which the syllabus and the examinations imposed upon his obvious enthusiasm for literature, Raleigh retreated into a jovial, off-hand style of lecturing which barely covered up his wounded pride:

I lecture in a very picaroon, jolly beggar, kind of way, think it wakes them up. On Crabbe I say: 'Why should we abuse Crabbe? He has never done us any harm: we have none of us read him.' On Keats I am tempted to say: '. . . you will none of you be any nearer Heaven ten years hence for having taken a B.A. degree, while for a love and understanding of Keats you may raise yourselves several inches. In any case, you cannot expect me to give you any facts about his life in one short hour. If you waste your time, I am determined not to waste mine.'[49]

Holding the Oxford Chair after 1904 meant less time spent lecturing, but Raleigh's contempt for institutionalized literary study only increased. In 1906 he wrote, 'I begin to hate criticism. Nothing can come of it.'[50] Seven years later this hidden anger had turned to complete cynicism, as he resolved simply to hang on to the money and keep quiet:

The worst of it is, I can't read Shakespeare any more, so I have to remember the old tags. Not that I think him a bad author, particularly, but I can't bear literature. This distaste must be watched, or they'll turn me out. It's their money I want, so I suppose I've got to go on and be an old mechanical hack on rusty wires, working up a stock enthusiasm for the boyish lingo of effusive gentlemen long since dead. I always said no good would come of poetry.[51]

A close examination of Raleigh's cynical reflections suggests that his problem was closely related to the conflict between the requirements of literary teaching and the 'personal' conception of literature and individual genius discussed above. This sonnet of Raleigh's ties them together:

I never cared for literature as such.
Iambic, dactyl, trochee, anapaest,
Do not excite my interest in the least;
And cultured persons do not please me much.
Great works may be composed in French or Dutch,
Yet my poor happiness is not increased.
To me the learned critic is a beast;
And poetry a decorated crutch.
    One book among the rest is dear to me.
'Tis when a man has tired himself in deed
Against the world, and falling back to write
Sated with love, or crazed with vanity,
Bemused with drink, or maimed by fortune's spite,
Sets down his Paternoster and his Creed.[52]

Here the root of his bitterness can be located in the loss of an ideal heroic literary authenticity — an ideal which recurs in his writings as often as his cynicism, and is in fact the reverse side of it. Raleigh always prized what he took to be the real human presence of an author over against the boring technicalities of literary works themselves. He often said that his

favourite books consisted of table-talk, letters, and autobiography; he would refer to Wordsworth, Blake, and Shakespeare as 'Bill'; and even when discussing books which he
disliked, he attacked the authors rather than the writings — as
for example when he dismissed what he called 'modernism'
out of hand because of the ugliness of Zola's and Ibsen's
portraits.

Raleigh's early book *Style* (1897) revealed a diehard
Romantic, declaring bluntly that the classical ideal stood
for paralysis and death. He followed Arnold in attributing
poetry's superiority over its rival 'the philosophic cripple' to
its concreteness: 'But literature, ambitious to touch life on all
its sides, distrusts the way of abstraction, and can hardly be
brought to abandon the point of view whence things are seen
in their immediate relation to the individual soul.'[53] Raleigh's
own distrust of the way of abstraction is expressed in a picture
of the world of scientific reason which it is hard to believe is
not self-parody:

There, words are fixed and dead, a botanical collection of colourless,
scentless, dried weeds, a *hortus siccus* of proper names, each individual
symbol poorly tethered to some single object or idea. No wind blows
through that garden, and no sun shines on it, to discompose the melan-
choly workers at their task of tying Latin labels on to withered sticks.
Definition and division are the watchwords of science, where art is all
for composition and creation.[54]

Criticism must overcome this world of death and bring about
resurrection, performing a magical rather than an analytic
function, as Raleigh implies by using the language of Prospero:
'the main business of Criticism, after all is not to legislate,
not to classify, but to raise the dead. Graves, at its command,
have waked their sleepers, oped, and let them forth. It is by
the creative power of this art that the living man is recon-
structed from the litter of blurred and fragmentary paper
documents that he has left to posterity.'[55]

Raleigh can be seen to have elevated the personal greatness
of the author to such a point that any interest in the litter of
paper documents seems sacrilegious. In fact his distrust of the
study of literary works reached a comparable hysterical level.
He asserted on one occasion that 'The eunuch was the first

modern critic'[56] — a point which he developed in this remarkable passage:

Bradley's book on Shakespeare is good. Of course it is not nearly gutsy enough but he gets there all the same. Even with it I can't help feeling that critical admiration for what another man has written is an emotion for spinsters. Shakes. didn't want it. Jerome K. Jerome is in some ways a far decenter writer than Brunetière or Saintsbury or any of the professed critics. He goes and begets a brat for himself, and doesn't pule about other people's amours. If I write an autobiography it shall be called 'Confessions of a Pimp'.[57]

Virginia Woolf remarked perceptively on the machismo element of self-assertion inherent in this attitude 'He was coming to feel that there is some close connexion between writing and fighting.'[58] As early as 1897 he could write that he had 'never known a soldier whose style was not ruined by dabbling in the poets',[59] and by the time the Great War broke out, he clearly preferred military heroism to his own unmanly duties. Abandoning literature, he put his pen at the service of the war effort. Dissatisfied with the lack of heroism involved in professing literature, Raleigh was attempting, as Woolf put it, to become a Professor of Life. His cynicism about teaching literature was not so much a question of gross professional misconduct (as Q. D. Leavis sees it),[60] but arose from a very clear-headed understanding of the ridiculousness of literary culture's ambitions for bringing about social change. For instance, this outburst against the powerlessness of an art tolerated by Philistine patrons could almost have been written by William Morris:

The art-lover never gets hold of the tiller. He is engaged in loving. I feel as if we were all dupes and fools, allowed to amuse ourselves by furthering art and taste while our masters get ready to spit and roast us. *They* will subscribe guineas, and glad to do it — keeps us quiet. And we are all eminent and influential — they don't mind, so long as we are kept out of the engine-room, and given beautiful little toy engines to play with in the saloon.[61]

Raleigh's failure lay in his refusal to overhaul his understanding of literature in this light. Instead he retreated to cultivating his lost ideal of heroic authorship and of a time when literature was supposedly really Life. This course did not increase literature's influence on the active world one iota: it simply bred a destructive literary self-contempt, so that when Raleigh's masters really did prepare to spit and roast all Europe in 1914 he supported them all the more unquestioningly.

At Cambridge, the establishment of an English school had taken a different, quieter course than at Oxford. At the setting-up in 1878 of a Board of Mediaeval and Modern Languages, English formed a sub-department of this larger body, from which it extricated itself step by uncontroversial step. Within the English School the bitter debates over philology which split Oxford were avoided, first by the provision of a specifically literary Chair in 1911, and then by the appointment in 1912 of H. M. Chadwick as Professor of Anglo-Saxon. Chadwick, himself originally a Classicist, was worried by the pedantic tendencies of his own subject, and insisted that Anglo-Saxon be studied along with the whole of ancient Northern European culture rather than acting as a dead weight on students of post-Chaucerian English literature. Accordingly, he agreed to break away and attach his department to the School of Archaeology and Anthropology (itself a product of the demands of colonial administration). This left the English School, when it was finally established on an independent basis in 1917, with a far greater literary bias than Oxford has ever conceded. The problems of teaching a subject with no appropriate factual (and above all examinable) staple were, if anything, only the greater because of this, though they did not express themselves nearly so violently as in Raleigh's letters.

The first King Edward VII Professor of English Literature, A. W. Verrall, died before he could take these problems on. His successor, Sir Arthur Quiller-Couch, was a surprising appointment. The terms under which the newspaper magnate Harold Harmsworth (later Rothermere) had established the Chair had left the appointment in the hands of the Crown, and Quiller-Couch, a patriotic Cornish squire, appears to have been chosen as a reward for services to the Liberal Party. Like Arnold, 'Q' was a 'Hellenist' Liberal who felt himself remote from the Nonconformist (and especially from the teetotal) traditions of mainstream Liberalism. He was not slow to announce his literary-critical credo, presenting it in his inaugural lecture in this form:

I propose next, then, that since our investigations will deal largely with style, that curiously personal thing; and since (as I have said) they cannot in their nature be readily brought to rule-of-thumb tests, and may therefore so easily be suspected of evading all tests, of being mere dilettantism; I propose (I say) that my pupils and I rebuke this suspicion by constantly aiming at the concrete, at the study of such definite beauties as we can see presented in print under our eyes; always seeking the author's intention, but eschewing, for the present at any rate, all general definitions and theories, through the sieve of which the particular achievement of genius is so apt to slip. And having excluded them at first in prudence, I make little doubt we shall go on to exclude them in pride. Definitions, formulae (some would add, creeds) have their use in any society in that they restrain the ordinary unintellectual man from making himself a public nuisance with his private opinions. But they go a very little way in helping the man who has a real sense of prose or verse. In other words, they are good discipline for some thyrsus-bearers, but the initiated have little use for them. As Thomas à Kempis 'would rather feel compunction than understand the definition thereof' so the initiated man will say of the 'Grand Style', for example — 'Why define it for me?' When Viola says simply . . . [there follows a page of touchstones of between one and four lines in length from Shakespeare, Milton, Gray and Keats] . . . 'why then (will say the initiated one), why worry me with any definition of the Grand Style in English, when here, and here, and again here — in all these lines, simple or intense, or exquisite or solemn — I recognise and feel the *thing*?'[62]

The hostility to scientific understanding, the restraint of public expression of opinion, the importance of the 'grand style' for the 'initiated' — this combination is not only recognizably Arnoldian, right down to the touchstones; it is clearly an approach to literary study which had undergone no conscious development since Arnold formulated it. These principles provided no practical guidelines to those who first had the task of teaching English literature in an extended and supposedly systematic way. Like so many of the early professors of English Literature, Quiller-Couch was a former Oxford 'Greats' man; and the only teaching he seems to have enjoyed was his series of classes on Aristotle, from which women were excluded. His solution to the problems of systematic teaching was to treat the subject as a branch of rhetoric: his series of lectures *On the Art of Writing* (1916) is an honest if pedestrian attempt to see 'how it is done' in literary composition. Like Raleigh, he tackled the problem of organized literary study by rejecting it in favour of the cult of heroic practice. And, like Raleigh, he began to identify writing and fighting as the virile alternatives to mere scientific study. In his Preface to *On the Art of Writing*, written in 1915, he states:

Literature is not a mere Science, to be studied; but an Art, to be practised. Great as is our own literature, we must consider it as a legacy to be improved. Any nation that potters with any glory of its past, as a thing dead and done for, is to that extent renegade. If that be granted, not all our pride in a Shakespeare can excuse the relaxation of an effort — however vain and hopeless — to better him, or some part of him. If, with all our native exemplars to give us courage, we persist in striving to write well, we can easily resign to other nations all the secondary fame to be picked up by commentators.

Recent history has strengthened, with passion and scorn, the faith in which I wrote the following pages.[63]

Arnold, and many of the theorists of adult education, had looked to the study of English literature as an agent of social harmony, capable of binding class to class; and this social project moulded the forms of literary education as they were applied variously to the teaching of workers, school-children, women, and Indians. English literature became, in these applications, an impressive cultural example, a museum of national genius, but very rarely an object of critical investigation. Just when it seemed to have been given the chance to become a proper subject of extended study at university level, its social mission came to the fore again: the binding of class to class in common respect for the national heritage and all that was precious in it, against the threat of its destruction by the barbaric Hun.

NOTES

1. *CPW* iii. 64 ('Tractatus Theologico-Politicus').
2. Vicesimus Knox, *Elegant Extracts* (1824 edn.), Preface.
3. H. G. Robinson, 'On the Use of English Classical Literature in the Work of Education', *Macmillan's Magazine*, II (1860), 427. Cited by Palmer.
4. Ibid., 430.
5. *The Muse Unchained* (1958), 9. One honourable exception – a sober account which does discern the real trends at work behind all the personalities – is D. J. Palmer's *The Rise of English Studies* (1965), to which some sections of this chapter are indebted.
6. *Oxford Magazine*, 4 May 1887. Cited by Palmer.
7. Cf. Nowell Smith, writing in 1917: 'The irruption of that turbulent rascal, natural science, has perhaps had most to do with humanising our humanistic studies.' 'The Place of Literature in Education', in A. C. Benson (ed.), *Cambridge Essays on Education* (Cambridge, 1917), 110.
8. *How to Teach English Literature* (1880), 2, 42.
9. 'The Teaching of English', in *Essays on a Liberal Education*, ed. F. W. Farrar (1867), 310. Cited by Palmer.
10. Robinson, 'English Classical Literature', 427.
11. J. Llewellyn Davies (ed.), *The Working Men's College 1854-1904* (1904), 182, 98.
12. 'Education', in *Speeches on Politics and Literature by Lord Macaulay* (1909), 358.
13. See J. C. F. Harrison, *A History of the Working Men's College 1854-1954* (1954), 82-5. Hughes, a keen pugilist, had shown his fighting spirit in the turbulent days of April 1848, when he enrolled as a special constable to defend the government against a threatened Chartist demonstration.
14. Lord Playfair, 'The Evolution of University Extension as Part of Popular Education' in R. D. Roberts (ed.), *Aspects of Modern Study* (1894), 8.
15. Davies (ed.), *Working Men's College*, 8.
16. J. C. Collins, *The Study of English Literature* (1891), 147-8. Cited by Palmer.
17. Robinson, 'English Classical Literature', 429.
18. Revd Canon Browne, 'The Future of University Extension in London', in Roberts (ed.), *Aspects of Modern Study*, 27.
19. Robinson, 'English Classical Literature', 431.
20. *Life of Thomas Cooper, Written by Himself* (1872), 63. Cited by Palmer.
21. Robinson, 'English Classical Literature', 431.
22. Rt. Hon. J. Morley, in Roberts (ed.), *Aspects of Modern Study*, 82-3. Cf. Arnold's declared aim in *Culture and Anarchy*: to make Protestant fanatics and popular rioters 'pause and reflect' (*CPW* v. 160).
23. D. Beale (ed.), *Reports Issued by the Schools Inquiry Commission on the Education of Girls* (1869), 145.

24. F. D. Maurice, 'Plan of a Female College for the Help of the Rich and the Poor', in *Lectures to Ladies on Practical Subjects* (1855), 14.

25. Ibid., 17.

26. Beale (ed.), *Reports on the Education of Girls*, 87. Even the curriculum of Queen's College, 'advanced' by the standards of its time, betrays a strong bias in this direction. See the appendix to Rosalie Glynn Grylls, *Queen's College 1848–1948* (1948), 113.

27. For the use of bowdlerized editions of Shakespeare and Sheridan, see Grylls, *Queen's College*, 66.

28. C. Kingsley, 'On English Literature', in *Literary and General Lectures and Essays* (1880), 259.

29. Ibid., 258–9. Cited by Palmer.

30. Beale (ed.), *Reports on the Education of Girls*, 2.

31. Civil Service of the East India Company, *Report to the Rt. Hon. Sir Charles Wood* (1855), 9. Cited by Palmer.

32. Ibid., 21.

33. *Speeches by Lord Macaulay*, 115.

34. Ibid., 124–5.

35. 'The Literature of Britain', 342.

36. *The Health Exhibition Literature Vol. XV: Conference on Education, Section C* (1884), 231.

37. Civil Service of the East India Company, *Report*, 6.

38. *Report of the Schools Inquiry Commission* (1868), v. 521.

39. Palmer points to the apparent paradox that English studies grew, especially at the London Colleges, from the two complementary middle-class movements — Utilitarianism and Evangelicalism — which have normally been associated (among followers of Arnold at least) with hostility to imaginative writing. Palmer, 15.

40. Collins, *Study of English Literature*, 62.

41. E. A. Freeman, 'Literature and Language', *Contemporary Review*, lii (1887), 562. Cited by Palmer.

42. T. Case, *An Appeal to the University of Oxford against the proposed Final School of Modern Languages* (Oxford, 1887). Cited by Palmer. The anti-Teutonism here may appear alarmist, but the provocation from the Teutonists in recent years had been just as remarkable. A notable case was the series of lectures given by Charles Kingsley as Professor of History at Cambridge, collected under the title *The Roman and the Teuton* (1877). The lectures aimed to justify the actions of 'our race' (the Teutonic) in destroying the Roman empire, claiming that 'the hosts of our forefathers were the hosts of God' (p. 306). Set against this kind of (literal) Vandalism, Case's fears of a reversion to the Dark Ages begin to make more sense than at first sight.

43. *The Times*, 6 December 1893. Cited by Palmer.

44. Beale (ed.), *Reports on the Education of Girls*, 145.

45. The importance of this view is reflected in the title of Maurice's collected essays, *The Friendship of Books* (1874); and again in *Shakspeare Personally* by David Masson (1822–1907), another pioneer of

women's education who was Professor of English Language and
Literature at University College, London, before taking up the English
Chair at Edinburgh.

46. *The Letters of Sir Walter Raleigh 1879-1922*, ed. Lady Raleigh
(1926), 532.

47. Ibid., 26.

48. Ibid., 122. Cited by Palmer.

49. Ibid., 125.

50. Ibid., 296.

51. Ibid., 396.

52. Ibid., 329.

53. Walter Raleigh, *Style* (1897), 59.

54. Ibid., 41.

55. Ibid., 128-9. Cf. *The Tempest*, v. i. 48-9.

56. Raleigh, *Letters*, 220.

57. Ibid., 268-9.

58. Virginia Woolf, 'Walter Raleigh', in *Collected Essays*, I (1966),
317. In *A Room of One's Own* (1929), she wrote: 'I began to envisage
an age to come of pure, of self-assertive virility, such as the letters of
professors (take Sir Walter Raleigh's letters, for instance) seem to fore-
bode, and the rulers of Italy have already brought into being' (p. 154).

59. Raleigh, *Letters*, 195.

60. Q. D. Leavis, 'The Discipline of Letters'. *Scrutiny*, xii (1943),
12-26.

61. Raleigh, *Letters*, 277.

62. A. Quiller-Couch, *On the Art of Writing* (Cambridge, 1916),
14-16.

63. Ibid., Preface.

4. LITERARY-CRITICAL CONSEQUENCES OF THE WAR

'I'm all for law and order, and hurra for a revolution!'
Thomas Hughes, *Tom Brown's Schooldays*

The First World War has long been recognized as marking a distinct turning-point not only in world history but also in English literature, giving rise to the literary renaissance headed by Joyce, Eliot, and Lawrence. What is often obscured is that the discipline which has arrived at this assessment — English literary criticism — owes its own renaissance largely to the same catastrophe. The inauguration of the modern epoch of wars and revolutions triggered, as one of its remote ripples, what one of its participants described as the 'Revolution in English Studies'.[1] This 'revolution' may have reached its maturity only with the launching of *Scrutiny* in 1932, but the historic breakthrough (as F. R. Leavis was later to see it[2]) for the ideals to which that journal appealed was made at the time of the war, and in its immediate aftermath. As Basil Willey was to recall, 'the most significant thing, the genuine revolution, was the War of Independence whereby English became an autonomous discipline, free from all alien tyrannies and ancient prejudices'.[3] The vocabulary here catches the particular wartime attitudes within which the newly independent discipline was hatched. The resurgence of national pride, and the indignant brandishing of the cultural heritage that went with it, acted as a powerful impetus to the establishment of English Literature as a 'central' discipline.

To this day, a literary–critical exercise widely used in schools is the comparison of the war poems of Rupert Brooke with those of Wilfred Owen: an exercise designed to celebrate the capacity of poetry to rise above the pressures of cliché and propaganda under the test of great events. Taken on its own, though, the contrast cannot convey just how extensive those pressures were upon English literature, or how singular Owen's honesty was. For the pressures were there, not just in the form of a certain romantic naivety detectable in Brooke's verses, but in the form of organized propaganda campaigns in the world of letters. From the beginning of the war until the fall of Asquith, the British government's War Propaganda Bureau under Cabinet minister C. F. G. Masterman had its own Literature and Art Department supported by newspaper magnates Beaverbrook and Rothermere, and with Anthony Hawkins (the 'Anthony Hope' of *The Prisoner of Zenda*) on its staff. A significant role in its efforts was played by the three leading novelists who had large readerships in the United States: John Galsworthy, H. G. Wells, and Arnold Bennett (who was to become Director of British Propaganda in France in 1918). The German population of the United States, which Matthew Arnold had mentioned specially as a potential vanguard of culture in that country, was to be neutralized by a stream of apologetic articles in the American press by Britain's foremost novelists; and after the Easter Rising of 1916, the American Irish were to be answered by the same means. But this triumvirate was only the advance guard of literary patriotism: on 18 September 1914, *The Times* published 'Britain's Destiny and Duty' (subtitled 'A Righteous War'), a statement in support of the British government's version of its war aims, over the signatures of no less than fifty-two celebrated writers of the day. These included H. Granville Barker, J. M. Barrie, A. C. Benson, A. C. Bradley, Robert Bridges, G. K. Chesterton, Arthur Conan Doyel (*sic*), H. Rider Haggard, Thomas Hardy, Rudyard Kipling, John Masefield, Gilbert Murray, Henry Newbolt, Arthur Pinero, Arthur Quiller-Couch, and G. M. Trevelyan. This was greeted as 'more than an outburst of unthinking patriotism' by a leading article the next day, which observed: 'The great virtue of the old German culture was that it was disinterested. It believed in truth and justice, not German truth or German justice. The vice of modern German culture is that it always has a German axe to grind. And that is the reason why the intellect of the world is deaf to its sophistries.'[4]

i. *The alien yoke*

The principle of 'disinterestedness' in English letters led to an increased pressure to throw off what Willey refers to as the 'alien yoke of Teutonic philology'. As was noted in previous chapters, the question of 'Teutonism' had fuelled many a controversy over syllabus in the movement for English studies. With the advent of war with Germany, though, the balance was tilted decisively against any respect for 'Teutonic' culture. John Burnet, Dean of the Arts Faculty at St. Andrews University, insisted in 1917 that 'It would be of little use to defeat the Germans in the field if we were to fall under the influence of German *Kultur*, and this danger is nowhere so great as in all matters connected with education.'[5] As early as 1914 F. S. Boas, a former Oxford extension lecturer and Professor of History and English Literature at Queen's College, Belfast, was remarking: '[Wordsworth's] disillusionment, which was shared by Coleridge and Southey, in the French as apostles of liberty has its parallel at the present time in the equally bitter disillusionment of those who have looked upon the Germans as apostles of culture and not of *Kultur*.'[6]

At Cambridge, Quiller-Couch was at the forefront of this campaign to denigrate German culture. He praised Matthew Arnold for calling a halt to a tide of Germanism in the nineteenth century, though even Arnold was guilty of over-valuing German literature.[7] As for German scholarship, Quiller-Couch argued, in a series of lectures entitled 'Patriotism in English Literature', that it was completely unfit to approach the beauties of English Literature: echoing, almost word for word, Raleigh's condemnation of science as the world of death, he said of the German authorities on English: 'this lovely and living art which they can never practise nor even see as an art, to them is, has been, must be for ever, a dead science — a *hortus siccus*; to be tabulated, not to be planted or watered'.[8] This incapacity was inevitable because, 'if only by the structure of his vocal organs a German is congenitally unable to read our poetry . . .'.[9] Quiller-Couch went so far that the *Cambridge Review* was obliged to reprimand him for lecturing too much on the deficiencies of German criticism and not enough on the qualities of English literature.

At Oxford Raleigh took the opportunity to declare that 'German University culture is mere evil',[10] and dreamed up a typically heroic alternative to the mass destruction of the war: 'I should like to get up a team of 100 Professors and challenge 100 Boche professors. Their deaths would be a benefit to the human race.'[11] As for the Boche professors who found themselves stranded in Britain, those who were not interned found themselves effectively disenfranchised within the influential university committees — even in the case of Professor Breul, a naturalized British citizen who lost a son on the British side in the war. The effective demotion of Breul and Braunholtz in the Cambridge Modern Languages School certainly seems to have facilitated the introduction of an English course virtually free of philology. This anti-Teutonism continued after the war, reinforcing the particular conceptions of literary study which the English School's founders held. Willey recalls that 'In the days of "Q"'s greatest influence and popularity, that is, the decade following the first great war, all things Teutonic were at a discount, and "Q" was able to score points off philology, off all accumulations of mere facts about literature . . . by calling them "Germanic".'[12] At Oxford too, Raleigh's successor George S. Gordon used the same patriotic device to introduce his own ideas of how English should be studied: 'The war, which broke so many things, cannot be considered as wholly malignant in its consequences if it should prove to have broken our servility to the lower forms of German scholarship, that nightmare of organised boredom by which all grace and simplicity and nature were frightened from our studies.'[13] Gordon went on to defend the 'intensely national' traditions of English scholarship against the tyranny of method and such continental barbarities as seminars.

These arguments were the response of the 'literary' faction of the English studies movement to the opportunity of pushing philology decisively into a subordinate position. But the patriotic fervour of the war years was used to gain a hearing for English studies as a whole, appealing to national pride in the language as well as the literature. Raleigh's wartime assertion that, now, the English language 'is safe to be the world language'[14] indicates the theme of this campaign. It continued in the Newbolt Report of 1921 on *The Teaching of English in England*: 'English is nearer than ever before to becoming a universally known language. The conditions created by the war have spread the knowledge of our language over the five continents of the earth.... Most of this extension of English may be due to political or commercial reasons. But there are higher reasons too. The intrinsic value of our literature is increasingly recognized.'[15] The report argued that a national consciousness of pride in the language, similar to that shown by the French, had emerged in Britain as a result of the war, and that, if cultivated by the study of English, it could provide the basis for a lasting national unity.

The war did not of course affect the defenders of English studies simply as an opportunity to advertise the importance of their work. Such a picture of these early pioneers as nothing more than educational war-profiteers would omit the fact that the catastrophe which gave them such golden opportunities was also shaking them up profoundly. English studies were undergoing a slow process of what has since been called 'professionalization', but this was not nearly far enough advanced by the time of the war for the movement to keep itself steady in such an acute crisis. The war had caught this literary movement groping forward with no clearly defined aims, no solid or cohesive institutions, and little to show for itself but bold pretensions which were now to be put prematurely to the test. This unexpected trial would sort out the cynics from the true believers, the amateurs in the loose sense from the amateurs proper, abruptly transforming the slow 'professionalization' of English studies into the basis for the emergence of a tested élite equipped to supervise the post-war literary renaissance.

The direct impact of the war upon the consciousness of the adolescent English studies movement cannot of course be directly weighed and measured. However, the contours of the problems which it posed can be discerned in many wartime lectures. An interesting case is this introduction to a lecture by the Shakespearian scholar C. F. E. Spurgeon, Professor of English Literature at University College, London:

I wished to make a little study of some purely literary theme, but each time I sat down to think about it, the moment my mind was freed from the necessary and merciful routine of a college lecturer's life, it seemed to fly back irresistibly like a released spring to some question of this kind, some question of ultimate values as seen afresh in the scorching and purifying light of war. . . . I think it may not be without some interest to try and record some of the impressions or reflections of a working teacher of literature, whose duty it has been to lecture on poetry during two years of the greatest conflict of material and spiritual forces the world has ever known.[16]

Spurgeon's reflections lead her, after initial doubts and soul-searchings at the beginning of the war, to a renewed conviction of poetry's importance, and indeed to a belief that the achievement of the sensibility required for poetry-reading is the main goal of life itself. She notes that the pressures of the war have encouraged a general 'heightening of sensation' fruitful for poetry; and the reaffirmative importance for her of such an abnormal state is carried over into her redefinition of poetry as above all a 'state of mind' or 'condition of being' rather than a set of written lines.

This keen sense of the war as a real test of the value of literary study, and the accompanying pressure towards viewing poetry in terms of ultimate spiritual values, was felt by a number of figures in the English studies movement, among them Ernest de Selincourt, another ex-Classicist formerly with the Oxford English School and by now Professor of English at the University of Birmingham. His wartime lectures, published as *English Poets and the National Ideal* (1916), began:

A time like the present, when we are in the throes of a great national crisis, affecting the lives of the most callous and indifferent of us, affords a clear test of the value that we really attach to literature, and, in particular, to poetry, the highest form of literature. Do we lay it aside as a pleasant pastime suitable enough for less hustling days but remote from our present practical needs and purposes, or do we turn to it with a keener spiritual hunger, feeling that it can give us not merely a pastime but in the true sense *recreation*?[17]

As in the case of Spurgeon's torments, de Selincourt's lectures bear witness to a fear of the literary student's social usefulness (indeed, loyalty) being cast in doubt. The 'keener spiritual hunger' for literature felt by these teachers carries with it a search for self-justification on a higher pitch. De Selincourt's lecture on Milton shows this anxiety rising close to the surface: after perpetrating a thinly disguised analogy between the Kaiser and Milton's Satan, he eludes the non-combatant's white feather by a vehement defence of Milton's example.

Of course there were those, both at the time and afterwards, who were
ready to pour contempt on his achievement and to taunt him because
he did not actually bear arms in defence of his cause. But he knew that
the service which he performed was most suited to his genius. . . . And
to that defence he gave *more* than the best twenty years of his life, the
prime of his manhood, for it was then that he became totally blind.[18]

The search for heroism of this kind, less disconcerting than
Raleigh's rejection of letters for war, is de Selincourt's major
preoccupation in these lectures. His tentative conclusion is
that the forging of a literary heritage is, in some way of its
own, a theatre of war. He refers, for example, to Shakespeare's
'capture by the enemy'[19] as good cause for a campaign on
this front, and he attempts to enlist even Byron, Shelley, and
William Morris into the camp of literary patriotism. Beyond
this retrospective recruitment campaign, though, lies an
appeal to the needs of the coming post-war reconstruction.
Having argued the case for literature to be considered an
important component of the national arsenal, de Selincourt
puts forward English studies' claim for a greater place in
peacetime too: 'At such a time we shall do well to live in
the companionship of the poets. . . . We can only avert
disaster by meeting the future in the spirit of the poet.'[20]
Closing his lectures with Shelley's resounding passage from
the *Defence of Poetry,* in which poets are acclaimed as the
'unacknowledged legislators of the world', de Selincourt
looks to the acknowledged legislators to recognize and
provide for the study of English literature as a potential bond
of national unity which can hold post-war society together.

ii. *The Newbolt Report*

Real legislators had learned enough from the war to respond
positively to pressure from educationalists. As long ago as
1866, the military victories of Prussia had been widely
attributed to her school system, and Matthew Arnold had
pointed out in his report two years later that the rate of
illiteracy among Prussian military recruits was 2 per cent, in
the British forces 57 per cent. Having learned it the hard way

the British state was not going to forget this lesson twice. One of its stated priorities in the immediate post-war period was to fortify this 'home front' of culture and education. As Lloyd George himself remarked in 1918: 'The most formidable institution we had to fight in Germany was not the arsenals of the Krupps or the yards in which they turned out submarines, but the schools of Germany. They were our most formidable competitors in business and our most terrible opponents in war. An educated man is a better worker, a more formidable warrior, and a better citizen. That was only half-comprehended before the war.'[21]

Within a year of the Armistice, the government had established four committees to investigate fully the state of teaching in the fields of Science, Classics, Modern Languages, and English. The patriotic poet Henry Newbolt was nominated to chair the committee established to inquire into the teaching of English. Newbolt it was who had composed the immortal lines:

> The Gatling's jammed and the Colonel dead,
> And the regiment blind with dust and smoke.
> The river of death has brimmed his banks,
> And England's far, and Honour a name,
> But the voice of a schoolboy rallies the ranks:
> 'Play up! play up! and play the game!'[22]

A poet who had so effectively popularized the identity of the national and the school spirits was a likely candidate to oversee the kind of educational renovation which Lloyd George had in mind. And Newbolt had a further factor in his favour: his presidency of an important educational pressure group, the English Association, which had been founded in 1906 'to promote the due recognition of English as an essential element in the national education.'[23] — a goal now considerably advanced by the introduction of an English paper into the public schools' common entrance examination in 1917. The Association's historian, Nowell Smith, was to attribute its success to its prestige in powerful circles, and more particularly to its banquets and the social amenities enjoyed at its lectures.

This approach seems to have paid dividends, since the

suggestion for the setting up of the government committee on English (which came from the Association) was met with a virtual blank cheque upon which the Association was free to inscribe its own aims. The committee, which included such figures as F. S. Boas, C. F. E. Spurgeon, C. H. Firth (Regius Professor of History), Quiller-Couch, and J. Dover Wilson, had a built-in majority of English Association members — nine out of fourteen. It was given terms of reference wide enough for it to propose rebuilding the entire 'arch' of national education around the 'keystone' of English.[24] Its report, published in 1921 as *The Teaching of English in England*, was greeted almost as a best seller (despite the public burning of a copy by Professor W. P. Ker of University College, London[25]), and became a guiding influence upon the development of English studies, particularly in the schools, but also in the universities through the work of I. A. Richards.

The committee acknowledged the debt owed by English studies to the war; they noted two tendencies arising from it whose convergence they could promote. First the new sense of national pride, for which literature was a standard-bearer: 'It is only quite lately that we in England have begun to have the definite consciousness, which the French gained in the age of Louis XIV, that we have a great and independent literature of our own, which need not lower its flag in the presence of the greatest on earth.'[26] The time had now come when English literature need no longer be seen largely from a German point of view, but could follow the example set by Britain's foremost wartime ally — that of cultural independence. The second important trend noted with satisfaction by the committee was the new importance attached by government and people alike to education. Not only were the majority of ordinary people, ex-soldiers in particular, clearly keen on education; in addition, their rulers had dropped the apprehension which many of them had shown in Arnold's day, now convinced that the populace could be trusted with the dangers of knowledge:

Yet in the great ordeal which the nation has just passed through, the schools, to a certain extent, came into their own. Many who had been inclined to discount them as factors in social and intellectual progress discovered them to be a power in the land. The discipline, adaptability, and intelligence manifested by the people at large made converts everywhere to the cause of education. Men serving with the forces revealed an unanticipated eagerness for instruction.[27]

The conjunction of a new national pride and a new recognition of the importance of education was expressed in the term often used in the Newbolt Report — 'national education'. It was the development of a system of education centred upon a national consciousness, based upon the native language and literature, which the committee took as its aim. Evidence of English studies' rightful pre-eminence in post-war education was already to hand in the statistics of extension lecture enrolments from 1913 to 1920, which showed a marked increase in the numbers studying English, both absolutely and in comparison with other available courses.

Such spontaneous confirmations of the importance of English, though, were not sufficient. They had to be built upon according to a certain plan of the new national education. Nor was national pride in itself sufficient unless it could be harnessed in a new unity between different social classes. Such a social reconciliation was the ambitious goal set for English studies by the Report's authors, under the slogan 'Culture unites classes' which Newbolt in his Introduction attributes to Matthew Arnold (a misquotation of *Culture and Anarchy* truer to Arnold's spirit than to his radical-looking letter: he had actually written that Culture 'seeks to do away with classes' [*CPW* v. 113], a very different objective). While the Classics tended to serve as an educational distinction between classes, a liberal education based on English 'would form a new element of national unity, linking together the mental life of all classes', and this applied to literature even more than to the common language: 'Such a feeling for our own native language would be a bond of union between classes, and would beget the right kind of national pride. Even more certainly should pride and joy in the national literature serve as such a bond.'[28]

The same strain is taken up with even more urgency in the Report's later section entitled 'Literature and the Nation', where it warns the government that this potential bond of unity between classes is currently too much loosened at the lower end:

The situation, as it was presented to us, is gloomy, though not entirely without the elements of hope. We were told that the working classes, especially those belonging to organised labour movements, were antagonistic to, and contemptuous of, literature, that they regarded it 'merely as an ornament, a polite accomplishment, a subject to be despised by really virile men.' Literature, in fact, seems to be classed by a large number of thinking working men with antimacassars, fish knives and other unintelligible and futile trivialities of 'middle-class culture', and, as a subject of instruction, is suspect as an attempt 'to sidetrack the working-class movement.' We regard the prevalence of such opinions as a serious matter, not merely because it means the alienation of an important section of the population from the 'confort' and 'mirthe' of literature, but chiefly because it points to a morbid condition of the body politic which if not taken in hand may be followed by lamentable consequences. For if literature be, as we believe, an embodiment of the best thoughts of the best minds, the most direct and lasting communication of experience by man to men, a fellowship which 'binds together by passion and knowledge the vast empire of human society, as it is spread over the whole earth, and over all time', then the nation of which a considerable portion rejects this means of grace, and which despises this great spiritual influence, must assuredly be heading for disaster.[29]

Even more ominously, it is revealed that the younger generation of workers especially 'see education mainly as something to equip them to fight their capitalistic enemies'.[30] The Report recognizes that great literature and 'common life' have become increasingly separated since the Middle Ages, and that this process has been accelerated greatly by the Industrial Revolution and the growth of the huge manufacturing cities (an analysis developed later by the Leavises). Hence the prevailing distrust of literature among the working classes. But at this point the Report suddenly casts aside its dismal warnings of growing class hostility, and invokes a romantic vision of the future poet, in a passage which concentrates the weight and direction of the committee's enthusiasm:

Here too lies our hope; since the time cannot be far distant when the poet, who 'follows wheresoever he can find an atmosphere of sensation in which to move his wings', will invade this vast new territory, and so once more bring sanctification and joy into the sphere of common life. It is not in man to hasten this consummation. The wind bloweth where it listeth. All we can do here is to draw attention to the existing divorce, and to suggest measures that may lead to reunion.

The interim, we feel, belongs chiefly to the professors of English literature. The rise of modern Universities has accredited an ambassador of poetry to every important capital of industrialism in the country, and upon his shoulders rests a responsibility greater we think than is yet generally recognised. The Professor of Literature in a University should be — and sometimes is, as we gladly recognise — a missionary in a more real and active sense than any of his colleagues. He has obligations not merely to the students who come to him to read for a degree, but still more towards the teeming population outside the University walls, most of whom have not so much as 'heard whether there be any Holy Ghost'. The fulfilment of these obligations means propaganda work, organisation and the building up of a staff of assistant missionaries. But first, and above all, it means a right attitude of mind, a conviction that literature and life are in fact inseparable, that literature is not just a subject for academic study, but one of the chief temples of the human spirit, in which all should worship.[31]

The framework of the argument here, which drives it on to such rhetorical conceits, is exactly congruous with Arnold's conception of Criticism's 'mission', in its concept of the *interim*. The historical scheme inherited from Saint-Simon and elaborated in *Essays in Criticism*, allotted a prominent role in human affairs to criticism, since the world was not yet ready for creative synthesis of its overwhelming new ideas and experiences. But certain significant contrasts with Arnold emerge more distinctly precisely at this point where this parallel historical conception is outlined. Where Arnold could exploit the impreciseness of the 'Criticism' to which the interim period belonged (guardedly refraining from proposing the establishment of an actual academy, for example) the authors of the Newbolt Report could not stop short of practical measures in this way, since they were already impli- cated in the real institutions of literary culture, bequeathed to them in part by Arnold. They could not place themselves so comfortably, so 'disinterestedly', at the starting-point of this intervening critical period when its first stages had already elapsed, and as a result they could not attain the same measured 'innocent language': their assertion that the interim belongs 'chiefly to the professors of English literature' has that same appearance of professional pleading which can be seen in the anxious wartime apologiae of de Selincourt and Spurgeon. Most of the committee's members were in fact 'professors of English literature' with the small 'p' appropriate to their religious terminology. And it is this (as with Spurgeon and de Selincourt) which gives rise to the unsteadiness of their rhetoric of Culture, in which a vocabulary of sanctification and worship jostles with the everyday professional 'staff', 'colleagues', 'degrees'. These overlaid religious references, on the one hand, and the practical urgency on the other, Arnold could afford not only to do without, but to dismiss as ugly committee-room Philistinism.

Immediately following this passage, indeed, the committee stresses the incompleteness of Arnold's message, arguing that it needs to be purged of any aloofness from common life, and that Culture must be 'propagated' in the smokiest of industrial centres. This new sense of urgency cannot be attributed solely to a kind of emerging caste outlook in the new profession looking for an expansion of its staff. It indicates a further pressure which had broken through even Arnold's rhetorical poise at crucial moments: the threat of monster demonstrations and working-class uprising.[32] The Newbolt Report's more deliberate examination of working-class attitudes to literature reflects a far more serious estimation of this problem than Arnold's throwaway remarks about the Populace; for now it seemed, with the formation of the Communist Party of Great Britain and trade-union powers like the Triple Alliance (both during the period of the committee's investigations), that there was more at risk than just a few railings in Hyde Park.

iii. *Immaterial communism*

The way in which such considerations encouraged the development of Arnoldian themes well beyond their original measured aloofness and into a new shrill tone of warning, can best be seen in the work of two of the committee's members: John Dover Wilson and George Sampson. Dover Wilson, who drafted the passage in the Newbolt Report quoted above on the impending disaster signalled by working-class alienation from literary culture, had a busy war working both for the Ministry of Munitions and as a school inspector for the Board of Education. It was his misfortune to have contributed to an early volume of wartime propaganda, *The War and Democracy* (1914), an apologetic chapter on Russia which attempted to play down the anti-Semitic pogroms and counter-revolutionary terror, presenting the 'misunderstood' Tsarist regime to British liberal opinion as a 'clean' ally behind which the loyal Russian people were firmly united, with the exception only of 'a small minority of theorists' preaching atheism and, worse still, 'free love'.[33] With the authority of what looked like first-hand knowledge (he had spent three years in Finland), Dover Wilson assured his readers that in conservative and Christian Russia they could firmly rule out the possibility of 'the social revolution of which Karl Marx dreamed'.[34]

The Russian empire for which he declared his love in this remarkable essay let Dover Wilson down very badly only three years later, in the shape of a veritable Marxist nightmare from which he never fully recovered. Even in a later and apparently 'strictly literary' work like *What Happens in Hamlet* (1935), he can be found referring gloomily to 'the communistic or corporative era that seems to lie immediately before us',[35] and the same shock certainly informs his sense of urgency in the Newbolt Report and further educational writings. Of these, the most pertinent in this context are his introductions to *The Schools of England* (1928) and to his edition of Arnold's *Culture and Anarchy* (1932); in both works he displays a fear of revolution enhanced by the recent General Strike. In the first, he argues that Matthew Arnold's hopes for the unifying power of humane education have been proved right by the marked decline in religious sectarianism since 1870. Before quoting Arnold's claim that culture seeks to do away with classes, he remarks:

Within two years of the General Strike is it idle to suggest that Arnold's remedy will prove as sovereign in the economic and social sphere as it has in the sectarian? Our industrial troubles arise, I am convinced, from the same cause as produced the denominational strife of previous generations; a whole section of the people feels itself to be disinherited, to be living and working outside the pale of privilege and opportunity. . . . The next step in the development of the English educational system . . . must inevitably bring with it some kind of alliance between culture and industry, to the great advantage of both, since what is wrong with our industry is not so much low wages or long hours as its lack of social meaning in the eyes of those performing its operations, and what is wrong with our culture is its divorce from the crafts of common life.[36]

If working-class children could be more effectively incorporated into the national literary culture, Dover Wilson seems to suggest, then they might put up with low wages and long hours in exchange for a purely educational 'equality'. The same suggestion is repeated almost word for word in his introduction to *Culture and Anarchy,* where the sense of threat looms even larger: 'Though the shadow of domestic anarchy under which it was written has to some extent passed away, if it be not too bold to say this within five years of the General Strike, a huger shadow has taken its place, that of a world-anarchy which threatens to bring the whole structure of civilisation toppling to the ground.'[37]

The same nervousness haunts the work of Dover Wilson's fellow committee-member George Sampson in his *English for the English* (1921). This book, in its concerns, its date of publication, and its subsequent influence, is virtually a companion volume to the Newbolt Report. It carries further the Report's twin developments of Arnold's vision of Culture — towards practical projects on the one hand, and towards a spiritual affirmation of these projects on the other. The practical programme of classroom exercises which it advocates draws its strength from Sampson's faith in the creative capacities of children — which has earned him a substantial respect as an educational theorist. But in his arguments for this approach to the teaching of English, Sampson introduces a more disturbing element — a compulsive reference to primary spiritual forces:

What the teacher has to consider is not the minds he can measure but the souls he can save. . . .

At this stage of our national education what matters is the faith, not the works. We have (so to speak) to undergo conversion, not to practise new austerities. . . .

The teacher's hardest struggle is not against pure ignorance but against evil knowledge. . . .

[English] is the one school subject in which we have to fight, not for a clear gain of knowledge, but for a precarious margin of advantage over powerful influences of evil. . . .

[The reading of English literature] is not a routine but a religion . . . it is almost sacramental.[38]

As a rousing finale, Sampson concludes the book by rounding on Classical studies — normally an ally of English against Germanism, but now a dispensable one:

How is the enemy's growing tyranny to be most effectively fought today — by the weapons of Shakespeare and Milton or of Sophocles and Homer? To escape the foul influence must we always go back to the old things and funny things that are foreign and not to the old things and funny things that are native? Is Prometheus less significant in Shelley than in Aeschylus? It is because I know that the power of the evil is so strong and the power of the good as yet so small that I beg the place of honour in the fight for our own great native force — 'the illustrious, cardinal, courtly and curial vernacular' of England.[39]

This is a more fiery rhetoric than that of the Newbolt Report. Not only does it incorporate the (mysterious) enemy of wartime usage, but its religious vocabulary, where that of the Report is ecclesiastical — referring to institutions and offices — is in Sampson's case more confessional, if not apocalyptic. It corresponds more closely to Arnold's picture, in *Culture and Anarchy,* of forces of darkness and light doing battle. Arnold, we have seen, had adroitly subsumed social conflicts into this contest of spiritual powers, of the 'best self' against the ordinary self. Sampson, following the same procedure, comes to conclusions almost identical with Arnold's in the latter's essay on 'Equality', where he had asserted that a community of humane manners was a community of equals, regardless of social inequalities.

The beginnings of a humane education here advocated will not involve a domestic revolution, or a rearrangement of the social system, or a new scale of moral values, or a preference of one sort of -ocracy or -ism to any other, or an upheaval of any sort. A humane education is a possession in which rich and poor can be equal without any disturbance to their material possessions. In a sense it means the abolition of poverty, for can a man be poor who possesses so much?[40]

Where Sampson develops Arnold's conception further is in his more urgent appreciation of the political power which the Populace may soon come to wield or shape. Arnold's social commentary started from a similar inkling, but Sampson uses it as the basis for an openly populist discourse: 'The safety of the world and the future of civilisation depend upon the character and intelligence of the multitude. Rulers of whatever name or rank may rise and fall; the ultimate power to make or mar the world will always be with the masses.'[41]

This may not sound like the voice of Arnold, but the motive is the same: to construct a world of Culture in which upheavals and disturbances can be dissolved. The difference is that an 'innocent language' such as Arnold had attempted for this purpose is now increasingly difficult to sustain; so that instead, the more disreputable Philistine language of radicalism is mimicked, the better to dissolve its dangers from within, by homeopathy. Sampson was to go further in this project (and in identifying the unspecified 'enemy' of his concluding passage) in the Preface to the 1925 edition of *English for the English*. Here he uses Arnold's arguments against the short-sighted policy of 'payment by results' in education, but with a new strain of spiritualized populism:

I believe that attempts to banish humanising and civilising elements from schools for working class children are vitally evil; and I also believe that those who declare that boys and girls going into 'labour' have no need of education, are more dangerous preachers of social disorder, and more insidious enemies of their country, than any deluded Hyde Park orator spouting under a red flag. Something called 'communism' (of the material kind) is the most feared or the most fancied of present-day panaceas; but few people seem to bother about the communism that, by a divine paradox, is real because it is immaterial. For just as the sun rises alike on the evil and the good, so joy and hope and affection and the other movements of the spirit come alike to the rich and the poor if we do not impede them.[42]

The concept of an 'immaterial communism' introduced here is at the centre of the post-war English studies 'crusade'.

In its eagerness to send missionaries into the industrial centres to win the hearts and minds of the workers, the movement consciously resembled real-life bolshevism, but (as has been said of the Labour Party) it actually owed more in this respect to Methodism, by virtue of this absolute distinction made between equality in spirit and the threatening evils of the material world: 'You get real communism and 'brotherhood of man' at a concert or a theatre or a cricket match. Listening to *The Magic Flute* or *The Mikado* we are all equal, and that is the only kind of equality worth seeking for. Deny to working class children any common share in the immaterial, and presently they will grow into the men who demand with menaces a communism of the material.'[43]

Without doubting Sampson's sincerely felt sense of threat or of panic, it is worth while to note that the communist threat invoked here and by others in the history of the English studies movement is also a vicarious threat aimed by the apostles of Culture at those who will not take them seriously. These warnings to the government by loyal citizens tend to take on the politely menacing aspect of a distinct 'lobby'. The more far-fetched variant of this development can be seen in a speech made in 1928 by Sir Henry Newbolt himself, as a presidential address to the English Association. The Association had by this time succeeded remarkably well: the important Hadow Report of 1926 on adolescent education, for example, had enshrined its aims in the recommendation that 'The chief object in the teaching of literature is the communication of zest . . .'.[44]

After the success of the Newbolt Report and the best-selling *Poems of To-day*, the Association had seen a steady growth in membership to a peak of about 7,000, crowned by the presidency in 1927 of no less a person than the Prime Minister and victor of the General Strike, Stanley Baldwin. To Newbolt, there seemed to be no limit to this growth of influence: 'I have been asking myself, and I now ask you, whether the true idea of an English Association, that which it was to be, has not been waiting to reveal itself at the fateful moment, the moment of a great national need, which no other agency is at the present time capable of satisfying . . .'.[45] After reviewing the progress made since the Report of 1921 and calling upon his audience to 'hope with me for a national fellowship in which it shall be possible for everyone to forget the existence of classes' (an aspiration common enough in the history of the English studies movement), Newbolt unveils his astonishing proposal for this fateful moment of national crisis: 'I ask you for help to devise and put in operation a scheme under which your Association would ally itself with all those men and women in every neighbourhood who have grasped the fact of today – the salient fact that the present form of society is wholly inadequate to find place for all who are able to create and worthy to enjoy an unembarrassed sense of national unity.'[46]

Without prejudicing the question whether a coup master-minded by the English Association would have altered the course of British history for better or for worse, it has to be acknowledged that this kind of rallying-cry has considerable overtones of monomania. It seems that the visionary enthus-iasm represented by the manifestos of 1921 had fostered a view of the importance of Culture's 'mission' out of all proportion to its actual or possible social consequences, culminating in a deranged bitterness.

The critics of this trend of thinking were to attribute the derangement, plausibly enough, to a kind of wartime shell-shock. George Gordon, whose response is the most significant for the development of English as an academic discipline, recalled the impact of the war upon the literary world in these terms: 'Words went down in price; a gloom settled on the fraternity; they felt they didn't matter, and some of them could hardly bear it. The good ones got over it more or less; the conceited ones sulked; but some have never recovered from the awful suspicion of that time that literature was futile.'[47] For Gordon, Raleigh's protégé and successor at Oxford, the futility of literature was not an 'awful suspicion' but almost an axiom – making the assertions of the Newbolt Report simply a hysterical conceit. Like Raleigh, Gordon would insist on the primacy of 'things' over 'mere words', and regarded most aspects of his own profession (lecturing especially) with contempt. While this standpoint could have debilitating effects upon an acute mind like Raleigh's, Gordon seems not to have found it so uncomfortable: his attitude, for example, to an offer to edit an expurgated Shakespeare as simply 'a good commercial venture'[48] can be seen as evidence of a cultivated cynicism about literature. But it was possibly only from such a standpoint of professional irresponsibility that the wild pretensions of the Newbolt Report could immediately be recognized and resisted. Gordon's inaugural lecture as the Merton Professor takes up the challenge vigorously, starting with a eulogy to Raleigh and his essentially gentlemanly approach to literature, then turning upon the new zealots represented on 'one of those Departmental Committees which came in with the Peace' with a biting scorn:

The literary doctrine must be worth examining which on its higher or University branches bears such very poor fruit. It is briefly this: that England is sick, and that English Literature must save it. The Churches (as I understand) having failed, and social remedies being slow, English Literature has now a triple function: still, I suppose, to delight and instruct us, but also, and above all, to save our souls and heal the State. . . . Literature, to these reformers, is everywhere a sacrament, a holy remedy, and apparently almost the only sacrament now left.[49]

In response to the Newbolt Report's anxieties about the workers' distrust of literature, Gordon replies that he can see the workers' point — literature, after all, is not everything. It was possibly these remarks which provoked Sampson to add his warnings about an unholy alliance of bolsheviks and educational élitists to later editions of *English for the English*. And indeed, Gordon's attitude to popular education was, from Sampson's (or Arnold's) point of view, dangerously reactionary. Of one of his servants, he once wrote: 'I hope the new maid is as illiterate and competent as ever. It would be a sad day if she took to reading at her age! That's how Socialists are made.'[50] But Gordon's more patrician approach to (or rather, retreat from) English studies was not wholly alien to the Arnoldian tradition; it was just that he valued high culture more for its evidence of a 'grand style' of living, and he was worried lest the tone of practical urgency in the Newbolt Report should disrupt the all-important poise of Culture's 'innocent language', unnecessarily implicating it in social controversy: 'I observe, and not only in this Report, the growth of a religious jargon about literature and literary genius, and I observe it with regret as an affront to life. We must be modest to be believed.[51]

Gordon's response to the Newbolt Report, his insistence that extravagant pretensions could only damage the prestige of English study, marks a distinction between English at Oxford and at Cambridge which had by now become hardened. In recoiling so pointedly from the conclusions arrived at by the Cambridge Professor and his fellow committee-members, Gordon confirmed the tendency, which Raleigh had already initiated in the Oxford school, away from holding any 'faith' in English literature. The further exploration of English studies' social functions and duties, its possibilities as a 'central' discipline, could now be carried out only in the more highly charged atmosphere of Cambridge.

NOTES

1. Tillyard, *Muse Unchained*, subtitle.

2. F. R. Leavis, *English Literature in Our Time and the University* (1969), 11.

3. B. Willey, *Cambridge and Other Memories* (1964), 23-4.

4. *The Times*, 19 September 1914, 9 ('English Authors and German Culture'). For a fuller account of this episode, and of literary propaganda, see D. G. Wright, 'The Great War, Government Propaganda, and English "Men of Letters" 1914-1918', *Literature and History*, no. 7 (Spring 1978), 70-100.

5. John Burnet, *Higher Education and the War* (1917), 2.

6. F. S. Boas, *Wordsworth's Patriotic Poems and Their Significance Today* (English Association pamphlet, 1914), 7.

7. A. Quiller-Couch, *Studies in Literature* (Cambridge, 1918), 231-45 ('Matthew Arnold').

8. Ibid., 314.

9. Ibid., 311.

10. Raleigh, *Letters*, 474. A striking echo of this sentiment is to be found in the contemporary writings even of such an avant-garde figure as Ezra Pound: 'People see no connection between 'philology' and the Junker . . . the 'university system' of Germany is evil . . . the whole method of this German and American higher education was, is, evil, a perversion.' Ezra Pound, 'Provincialism the Enemy', *New Age*, 12 July 1917; rpt. in *Selected Prose 1909-1965*, ed. William Cookson (1973), 161.

11. Ibid., 457.

12. Willey, *Cambridge*, 7.

13. G. S. Gordon, *The Discipline of Letters* (1923), 20.

14. Raleigh, *Letters*, 468.

15. *The Teaching of English in England* (HMSO, 1921), 200.

16. C. F. E. Spurgeon, *Poetry in the Light of War* (English Association pamphlet 1917), 3.

17. E. de Selincourt, *English Poets and the National Ideal* (1916), 7.

18. Ibid., 58.

19. Ibid., 13. Quiller-Couch (*Studies in Literature*, 316-17), quotes 'prologues' inserted into wartime German productions of Shakespeare, welcoming him to his 'second home' after his abandonment by the fickle British. De Selincourt attributes the Germans' admiration for Shakespeare to the anti-democratic political world of the plays.

20. Ibid., 117.

21. Cited in M. Mathieson, *The Preachers of Culture* (1975), 70.

22. Henry Newbolt, *The Island Race* (1907), 87 ('Vitaï Lampada').

23. Constitution of the English Association, cited in Nowell Smith, *The Origin and History of the Association* (English Association pamphlet 1942), 5.

24. *The Teaching of English in England*, 5.

25. For this and an account of the proceedings of the Newbolt Committee, see J. Dover Wilson's autobiography, *Milestones on the Dover Road* (1969), 95-100.

26. Ibid., 198.

27. Ibid., 58.

28. Ibid., 22.

29. Ibid., 252-3.

30. Ibid., 254. That working-class demands for education were more utilitarian and political than their educators wished is confirmed in, for example, Harrison, *History of the Working Men's College 1854-1954* 159-60.

31. Ibid., 259.

32. '"As for rioting, the old Roman way of dealing with *that* is always the right one; flog the rank and file, and fling the ringleaders from the Tarpeian Rock!" And this opinion we can never forsake, however our Liberal friends may think a little rioting, and what they call popular demonstrations, useful sometimes to their own interests . . .'. [*CPW* v. 526, 223.] This indiscretion in *Culture and Anarchy* was suppressed after the first edition.

33. R. W. Seton-Watson, J. Dover Wilson, Alfred E. Zimmern, and Arthur Greenwood, *The War and Democracy* (1914), 182. I am indebted to Terence Hawkes for stressing to me the significance of Dover Wilson's writings.

34. Ibid., 176.

35. J. Dover Wilson, *What Happens in Hamlet* (Cambridge 1935), 28.

36. J. Dover Wilson (ed.), *The Schools of England: A Study in Renaissance* (1928), 25-6. In this introduction, Dover Wilson refers to

Ernest Barker's *National Character and the Factors in its Formation* (1927). Barker was a colleague of his at King's College, London, and it seems from the Arnoldian formulations in the chapters on education and literature that Dover Wilson may have had a hand in drafting it; for example: 'The more the literary tradition of a nation becomes a common content of the minds of its members, through the diffusion of education, the more is that nation united, and the more homogeneous is its life. What divides a nation internally may be even more differences in culture than economic differences.' (p. 222.)

37. J. Dover Wilson, Introduction to Matthew Arnold, *Culture and Anarchy*, (Cambridge, 1932), 28.

38. G. Sampson, *English for the English* (1921), 19; 33; 43; 59; 105. Sampson was a member of the Cambridge Advisory Council on Religious Instruction.

39. Ibid., 141.

40. Ibid., 138.

41. Ibid., 34.

42. G. Sampson, *English for the English* (1925 edn.), xiv.

43. Ibid., xv.

44. Board of Education, *Report of the Consultative Committee on The Education of the Adolescent* (1926), 193.

45. Sir Henry Newbolt, *The Idea of an English Association* (English Association pamphlet 1928), 4.

46. Ibid., 13.

47. Cited in M. C. Gordon, *The Life of George S. Gordon 1881–1942* (1945), 164.

48. *The Letters of George S. Gordon 1902–1942* ed. M. C. Gordon (1943), 23.

49. Gordon, *Discipline of Letters*, 12-13.

50. Gordon, *Letters*, 44.

51. Gordon, *Discipline of Letters*, 13.

5. 'ON THE SIDE OF THE ARTIST': T. S. ELIOT'S EARLY CRITICISM 1917–1924

'I seek an inheritance, incorruptible, undefiled, and
that fadeth not away; and it is laid up in heaven, and
fast there, to be bestowed at the time appointed, on
them that diligently seek it. Read it so, if you will, in
my book.'

Bunyan, *The Pilgrim's Progress*

In studying the ideas of major critics on the social function
of literary criticism, it might seem in the case of such a figure
as T. S. Eliot that the more overtly social, political, and
theological writings which followed his conversion to the
Anglican Church in 1927 are the most fruitful fields of
investigation; and these texts are indeed of great interest in
this connection. But the very directness of affiliation in these
later works, which offers them so readily as candidates for
this kind of investigation, at the same time disqualifies them
by comparison with his earlier work. This is the Arnoldian
dilemma re-enacted: for Eliot's influence as a critic is more
or less in inverse proportion to the extent of his identification
with particular social, political, and theological camps in a
given work. It is undoubtedly his earlier critical writings —
the occasional reviews, the collection *Homage to John Dryden*
(1924), and above all *The Sacred Wood* (1920) — which
commanded and still command the greatest attention from
teachers and students of literature. By contrast, *For Lancelot
Andrewes* (1928), in which Eliot declared himself a 'classicist
in literature, royalist in politics and anglo-catholic in religion',[1]
and most subsequent works for at least a decade were greeted
by large numbers of his former followers with a new circum-
spection, if not positive suspicion. Eliot had by then over-
played his hand, and was plainly no longer speaking the
'innocent language' which Arnold had aspired to.

The richly suggestive formulations of Eliot's earlier writings,
though, could be and were drawn upon by many outside
Eliot's doctrinal camp. This is not to erect a distinction

between a 'disinterested' period before 1928 and a 'committed' or partisan period after that date; first, because there was a gradual movement in Eliot's writings towards explicit social, political, and religious partisanship (certainly from 1923 onwards) of which the 1928 credo was only a culmination; and, more importantly, the commitments made public in 1928 were to a considerable extent embraced even before some of Eliot's earliest critical writing, as recent research into Eliot's early work has suggested. As early as 1916, the Oxford University Extension Delegacy published Eliot's *Syllabus of a Course of Six Lectures on Modern French Literature*, in which the second lecture is described in terms of the reaction against Rousseau and Romanticism, the return to classical restraint, and the doctrine of original sin. Eliot writes that 'The present-day movement is partly a return to the ideals of the seventeenth century. A classicist in art and literature will therefore be likely to adhere to a monarchical form of government, and to the Catholic Church.'[2]

A little hindsight can read most of Eliot's future development into that one short statement, the second sentence of which bears an uncanny resemblance to the credo published twelve years later. On this and other evidence[3] it seems likely that T. E. Hulme's anti-Romantic, anti-humanist doctrines of classical order and original sin were familiar, and attractive, to Eliot several years before the posthumous publication of Hulme's *Speculations* in 1924. The often breath-taking assurance of Eliot's early critical writings draws on the solid self-consistency of Hulme's consciously reactionary positions, as spelt out in Eliot's notorious triad of allegiances. This strength is matched by a corresponding more-English-than-the-English understatement and conceal-ment in Eliot's tone: a real model of the same classical restraint which Eliot preached as a literary ideal. As an occasional reviewer of books, Eliot had neither the leisure nor the licence to spell out the connections between his social and literary doctrines at any length; nor, at first, had he the inclination — as Ezra Pound's nickname for him, 'Old Possum', suggests. For a combination of reasons, then, the larger implications of Eliot's thought are more deeply submerged in his earlier writings, and they are all the more fertile in their influence for it.

The achievement of restraint, of an authoritative and objective tone in Eliot's early criticism, marks an implicit rejection of the increasingly over-inflated humanist rhetoric exemplified in the Newbolt Report — much as George Gordon was to criticize the Report for its immodesty. Against the strident and religiose Englishness of this kind of literary discourse, Eliot reached back to Arnold's European scope for a check to this unruly chauvinist trend. While the war was still in progress and English literary jingoism in full bloom, Eliot wrote: 'It is the final perfection, the consummation of an American to become, not an Englishman, but a European — something which no born European, no person of any European nationality, can become.'[4] In a literary milieu plagued by English patriotism, the announcement of such a paradoxical ambition displayed a far-sighted awareness of cultural development, which was to put its author at an immediate advantage in post-war literary life: while others were blinded by chauvinism, Eliot, at the right time, attempted to see European culture steadily and to see it whole. When the military conflict came to a halt and the accompanying cultural patriotism lost its immediate pretext, it was to become gradually apparent that foreigners (Pound, Joyce, Yeats, and Eliot) had by different means come to dislodge the native literary avant-garde. Eliot and Pound were to be hounded for this as 'literary bolsheviks' — hardly an accurate description of their politics, but a striking image of the way in which they seemed consciously to have seized the opportunity provided by the war for their own disloyal motives. In Eliot's case this calculating perception of literary opportunities certainly owed much to his expatriate status and the scope of cultural vision which it afforded. Both the early poetry — in particular *The Waste Land* — and the early criticism astonished readers in the 1920s by their casual European (and even Indo-European) range of allusion. This has since been criticized as a typically American 'tourist-erudition'.[5] But if Eliot's knowledge of European culture was broad rather than thorough, this was at the time the most appropriate response to the challenge of the war and its aftermath: a creative and

critical vision which did not attempt to measure up to the
international scale of the holocaust and its significance would
not have been, in Arnold's terms, 'adequate'. As Arnold had
insisted against the 'inadequate' Romantics, 'not deep the
poet sees but wide'.

It was to Arnold's achievement that Eliot appealed in
introducing his first and most celebrated volume of criticism,
The Sacred Wood. True to his later statement that 'the
majority of critics can be expected only to parrot the opinions
of the last master of criticism',[6] Eliot identified Arnold, easily
enough, as the master, but refused to be overawed. The title
The Sacred Wood, as George Watson has pointed out, refers
to a sacred grove mentioned in Frazer's *The Golden Bough,*
of which the priesthood is secured only by murdering the
previous incumbent.[7] Eliot's critical project at this period
was to finish off Arnold's long reign. In this combat he
certainly had advantages: his expatriate vantage-point; the
benefits of the Symbolist experiment in poetry, which
offered him answers to poetic problems Arnold had found
intractable; and a far more rigorous training in Idealist
philosophy. Yet Eliot's victory in wresting away the priesthood and defending it was achieved in great measure by his
daring use of weapons stolen in true guerrilla fashion from
Arnold himself. For Eliot, this was the truest form of homage
to the master: in his early criticism he rejects the commonplace attitude of reverence for writers of the past, insisting
instead that they are better served by being used or stolen
from. As Eliot wrote in one of his earliest published essays:
'The 'influence' of James hardly matters: to be influenced
by a writer is to have a chance inspiration from him; or to
take what one wants . . .'.[8] The same piece on James contains
one of Eliot's most famous thefts from Arnold — a reformulation of his 'intellectual deliverance': 'James's critical genius
comes out most tellingly in his mastery over, his baffling
escape from, Ideas; a mastery and an escape which are
perhaps the last test of a superior intelligence. He had a mind
so fine that no idea could violate it.'[9] Ideas, which Eliot
compares with rabbits running wild, seem not to be the only
threat to the finely balanced sensibility. In another early
essay he asserts that 'the forces of deterioration are a large

crawling mass, and the forces of development half a dozen men'. As part of this cultural revelation he even claims to have 'seen the forces of Death with Mr. Chesterton at their head upon a white horse'.[10]

It might seem this is the work of a critic more Arnoldian than Arnold. Even Eliot's first appraisal of John Donne credits the poet with 'seeing the thing as it really is'.[11] But there is in Eliot's early criticism a distinctive ingredient which distances him from such a complete orthodoxy, indeed makes him at times openly hostile to the master. This factor is Eliot's own poetical aspirations. Unlike Arnold, whose criticism and poetry were almost chronologically distinct, Eliot was still a practising poet as yet unacclaimed. Accordingly it is the issue of the relations between criticism and creation which at first most clearly separates him from Arnold, standing out as a constant object of enquiry and reformulation in his early criticism.

i. *Criticism and creation*

The emergence of this theme of criticism and creation in Eliot's writings is at first obscure. There is an assertion that James's criticism is 'in a very high sense creative'.[12] There is an article deploring British slackness and amateurism in letters, attributing this decadence to unspecified 'mixed motives'.[13] There is another assertion that the critic's tasks are those of comparison and analysis while judgement and appreciation are for lazy minds only. More helpfully, there is praise for the criticism of Ezra Pound, because his critical writings 'are the comments of a practitioner upon his own and related arts',[14] a mode of criticism which had been out of favour since Dryden's time. These hints do not emerge clearly from between the lines of Eliot's criticism until the publication of *The Sacred Wood* and the essay entitled (after Arnold again) 'The Function of Criticism' (1923).

Significantly, it is to Arnold's first series of *Essays in Criticism* rather than to the more popular second series that Eliot refers in 'making amends' to Arnold in the Introduction to *The Sacred Wood*. It was in this first series that Arnold had emphasized the immaturity of the Romantic poets, attributing it to the lack of a current of ideas in society capable of nourishing their creative power. This concern of Arnold's early, more 'classicist' criticism indicates in Eliot's opinion 'the centre of interest and activity of the critical intelligence' [SW, xii]. But this central interest in the relations between creation and its cultural preconditions (including critical traditions) can lead the critic to being lured out of purely literary concerns: 'The temptation, to any man who is interested in ideas and primarily in literature, to put literature into the corner until he cleaned up the whole country first, is almost irresistible.' [SW, xiii.] In the light of Eliot's subsequent trajectory, this statement may be read more as a confession than as a warning; but at this stage he is determined not to repeat what he sees as Arnold's error of overstretching criticism's scope. Indeed, Eliot begins to recommend greater restrictions upon the ambitions of the critic, asserting that many of those who have been tempted into criticism should rather use their critical ability to improve their own creative work. He insists that criticism and poetic creation are complementary and strongly interdependent activities. 'When one creative mind is better than another,' he remarks, 'the reason often is that the better is the more critical.' [SW, xiv.] But when the critical faculty breaks loose from its task of sifting and correcting within the creative process, it will tend to fall victim to a 'suppressed creative wish' [SW, 7]. 'Modern criticism is degenerate', as Eliot puts it in the essay 'The Perfect Critic', because it confuses two possible responses to a work of art, analysis or further creation; so failing to resolve the emotions aroused by the work in the direction of either. In this way 'impure desires' find no outlet, and the response to art remains confined to arbitrary sentimental associations.

Eliot's goal, on the other hand, recalls Arnold's norm of objectivity: 'The end of the enjoyment of poetry is a pure contemplation from which all the accidents of personal emotion are removed; thus we aim to see the object as it really is . . .' [SW, 14–15]. Again, though, the obviously Arnoldian formulation here is to be subjected by Eliot to a subtle but important qualification. The critical stance aspired to may be a form of objectivity, but it is not Arnold's 'disinterestedness' — which, as Eliot indicates in the essay 'Imperfect Critics', could only be an ideal. For, in Eliot's opinion, the critic has necessarily to be 'interested' — in artistic practice. The earlier approval of Pound's role as a practitioner critic is reiterated in *The Sacred Wood* in terms which make this professional partisanship quite ' clear. Coleridge is praised for 'writing as a professional with his eye on the technique' [*SW*, 19], and even non-practitioners are judged in similar terms: 'But Mr More and Mr Babbitt, whatever their actual tastes, and although they are not primarily occupied with art, are on the side of the artist. And the side of the artist is not the side which in England is often associated with critical writing.' [*SW*, 42.] This partisanship is included in Eliot's stipulations for the ideal critic, who would need to have 'a creative interest, a focus upon the immediate future. The important critic is the person who is absorbed in the present problems of art, and who wishes to bring the forces of the past to bear upon the solution of these problems.' [*SW*, 37-8.] What Eliot means by bringing to bear the forces of the past will be examined below; but for the moment the important point to be noted is the way in which Eliot, in *The Sacred Wood,* attempts to tie criticism much more closely to the immediate concerns of creative practitioners, seeking to subordinate it to these concerns and not to allow the critic's own tastes or talents to take priority.

The most impressive statement of this doctrine comes in the essay on 'Hamlet and his Problems', in which the classical 'objective' ideal is put forward in terms very close to those of Arnold's self-criticism in the 1853 Preface to *Poems*: 'In the character Hamlet it is the buffoonery of an emotion which can find no outlet in action; in the dramatist it is the buffoonery of an emotion which he cannot express in art.' [*SW*, 102.] What is extraordinary in this essay is not so much the doctrine of artistic 'escape from emotion', which is more cogently presented in 'Tradition and the Individual Talent', but the way in which the almost psychoanalytic concept of the displacement of creative energy is applied not only to the character and its author but to its *critics*: 'And Hamlet the character has had an especial temptation for that most dangerous type of critic: the critic with a mind which is naturally of the creative order, but which through some weakness in creative power exercises itself in criticism instead. These minds often find in Hamlet a vicarious existence for their own artistic realization.' [*SW*, 95.] The effect of this extension is to cast suspicion upon the motives of those critics who are not primarily creative artists in their own right. As Eliot puts it in 'The Perfect Critic', the artist is 'oftenest to be depended upon as a critic; his criticism will be criticism, and not the satisfaction of a suppressed creative wish' [*SW*, 7].

Eliot's motives for this campaign against the 'impure desires' of other critics would seem, at this stage, to be explicable largely in terms of his own ambitions as a practising poet. But as his discussion of the relations between criticism and creation develops, it appears that there is a more pressing concern at work in his arguments on this subject. Only three years after the publication of *The Sacred Wood*, Eliot was to refine his position in an essay of considerable importance — 'The Function of Criticism'. Here he gives his most systematic exposition of the matter:

Matthew Arnold distinguishes far too bluntly, it seems to me, between the two activities: he overlooks the capital importance of criticism in the work of creation itself. Probably, indeed, the larger part of the labour of an author in composing his work is critical labour; the labour of sifting, combining, constructing, expunging, correcting, testing: this frightful toil is as much critical as creative. I maintain even that the criticism employed by a trained and skilled writer on his own work is the most vital, the highest kind of criticism; and (as I think I have said before) that some creative writers are superior to others solely because their critical faculty is superior. There is a tendency, and I think it is a whiggery tendency, to decry this critical toil of the artist; to propound the thesis that the great artist is an unconscious artist, unconsciously inscribing on his banner the words Muddle Through.
· · · · · ·
But this affirmation recoils upon us. If so large a part of creation is really criticism, is not a large part of what is called 'critical writing' really creative? If so, is there not creative criticism in the ordinary sense? The answer seems to be, that there is no equation. I have assumed as axiomatic that a creation, a work of art, is autotelic; and that criticism, by definition, is *about* something other than itself. Hence you cannot fuse creation with criticism as you can fuse criticism with creation. The

critical activity finds its highest, its true fulfilment in a kind of union
with creation in the labour of the artist.

. . . . . .

   At one time I was inclined to take the extreme position that the
only critics worth reading were the critics who practised, and practised
well, the art of which they wrote. But I had to stretch this frame to
make some important inclusions ... And the most important qualification
which I have been able to find, which accounts for the peculiar importance of the criticism of practitioners, is that a critic must have a very
highly developed sense of fact. [*SE*, 29–31.]

This passage is important, first for clarifying Eliot's notion
of the asymmetrical relation between creation and criticism;
and secondly, in that its revision of the chief requirement
of the ideal critic reveals a concern wider than the
practising poet's self-protection. Whereas the old criterion
of practitionership (insisted upon by Ezra Pound) could
easily be interpreted as narrowly self-interested, the new
criterion — the 'sense of fact' — reflects Eliot's concern to
maintain a basis for authoritative and objective standards.
This strengthened commitment to objectivity and authority
before authorship is confirmed in the rest of the essay, but
in a manner which draws Eliot into controversy about
Romanticism, 'whiggery', religion, and society — precisely
the 'temptation' which he had scrupulously skirted in *The
Sacred Wood*. Eliot seems by now to be following Arnold's
critical trajectory almost step by step: from the problems of
poets and the examples available to them, to the social
necessity of critical standards. Indeed, in the sustained polemic
against J. M. Murry which takes up the first part of the essay,
Eliot explicitly equates Murry's 'Inner Voice' with Arnold's
'doing as one likes', and Arnold is imitated even in the strange
practice of picking on railway travellers as targets for spiritual
contempt: 'The possessors of the inner voice ride ten in a
compartment to a football match at Swansea, listening to the
inner voice, which breathes the eternal message of vanity,
fear, and lust.'[15]
   In the same year, in an essay on 'Marie Lloyd', Eliot ventured
even deeper into the political end of cultural controversy,
claiming that the middle class was 'morally corrupt' and that
'with the encroachment of the cheap and rapid-breeding

cinema, the lower classes will tend to drop into the same state of protoplasm as the bourgeoisie'.[16] Such comments would not seem out of place in some of Eliot's writings of the 1930s — and it might seem that a sudden transformation had taken place in 1923, breaking Eliot away from the restrained and objective stance of *The Sacred Wood*. But as we have noted above, there is greater continuity than this in Eliot's social concerns: he was later to introduce his most notoriously fanatical contribution to cultural and political debate — *After Strange Gods* (1934) — as a continuation of views first expounded in 'Tradition and the Individual Talent'.[17] The continuity inheres in the central 'problem of order'; and the essay on 'The Function of Criticism' in which this phrase is introduced provides the crucial link between its literary and social references. The meticulous specification of the interdependence of critical and creative writing is loaded with a much greater importance by the background of threatened cultural disintegration; and an apparently pedantic problem which had so troubled the author of *The Sacred Wood* seems almost to be given a new clarity and purpose by its polemical setting.

Oddly, Eliot's campaign to restrict criticism's pretensions, to make it 'objective', seems to require for its proper emphasis a wide cultural and social frame of reference which begins to threaten its 'objectivity' all over again. There is no doubt that the need to stabilize critical discourse, to prevent it from overstretching itself, is one of the foremost concerns of Eliot's early criticism. As he wrote in the essay on Massinger in *The Sacred Wood*:

English criticism is inclined to argue or persuade rather than to state; and, instead of forcing the subject to expose himself, these critics have left in their work an undissolved residuum of their own good taste, which, however impeccable, is something that requires our faith. . . .
  It is difficult — it is perhaps the supreme difficulty of criticism — to make the facts generalize themselves. . . . [*SW*, 123-4.]

In arguing for a criticism without opinionated controversy, judgement, or appreciation, Eliot is carrying further Arnold's remark that the critic's job is to 'get out of the way' of the work. The noisy voice of personal opinion is to fade away before the all-important creative work, and literary judgement is to form itself invisibly, as a spontaneously generated 'fact'. But it is only by being tempted further into semi-political quarrels that Eliot seems to be able to stress the dangers of any alternative approach — thereby risking embroilment in the very partisanship which he seeks to render invisible.

ii. *Simultaneous order*

It might seem strange that Eliot should concentrate so much attention upon the distinction between critical and creative functions. But in the context of Matthew Arnold's influence upon English criticism, the project can be understood. Arnold's much-debated assertion that poetry is a 'criticism of life' had, in Eliot's view, to be challenged. It threatened to dissolve what for Eliot as a practising Symbolist poet was an essential principle: the autonomy of the poetic artefact. Conversely and simultaneously it threatened to devalue the content of the word 'criticism' by broadening it to include all manner of literary references to the world. Eliot was therefore determined to re-establish the distinction between creation and criticism. But the problem was complicated still further by the fact that Arnold had, in 'The Function of Criticism at the Present Time', introduced into his scheme of the relations between criticism and creation a particular historical dimension involving alternating periods in which each power would predominate. So while confusing the capacities of poetic creation and criticism, Arnold had tended to separate them in time. Eliot's concern to subordinate criticism to the present problems of poetry, concentrated in the ideal of the practitioner critic, necessarily involved in attack upon this historical model of Arnold's:

It is . . . fatuous to assume that there are ages of criticism and ages of creativeness, as if by plunging ourselves into intellectual darkness we were in better hopes of finding spiritual light. The two directions of sensibility are complementary; and as sensibility is rare, unpopular, and desirable, it is to be expected that the critic and the creative artist should frequently be the same person. [*SW*, 16.]

Eliot is insisting here, not only that criticism and creation are complementary, but that they need to be in some way *simultaneous.* His differences with Arnold over the historical relation of the two activities converge upon a fundamental difference over the concept of history itself. Against Arnold's liberal humanist or 'Romantic' belief in a better future for society and for poetry, Eliot counterposes an essentially static, or 'classical' view of history and tradition, referring in 'Tradition and the Individual Talent' to 'the obvious fact that art never improves' [*SW*, 51], and putting forward a version of 'tradition' from which any idea of development, even of temporal sequence, is excluded:

. . . the historical sense involves a perception not only of the pastness of the past, but of its presence; the historical sense compels a man to write not merely with his own generation in his bones, but with a feeling that the whole of the literature of Europe from Homer and within it the whole of the literature of his own country has a simultaneous existence and composes a simultaneous order. This historical sense, which is a sense of the timeless as well as of the temporal and of the timeless and the temporal together, is what makes a writer traditional.

. . . what happens when a new work of art is created is something that happens simultaneously to all the works of art which preceded it.[18]

The word 'order' has often been singled out as the centre-piece of Eliot's literary and social criticism, but its exact significance is lost without the 'simultaneous'. A peculiar revulsion from the idea of history as a process or temporal movement seems to be essential to Eliot's order.

An indication of the particular concern with a 'timeless' tradition which inspired this kind of formulation can be found in the early essay 'Reflections on Vers Libre' (1917). Here Eliot discusses the social prerequisites of poetic creation, and maintains that only in a 'closely-knit and homogeneous society' can certain poetic forms be carried to perfection. Moreover, such societies can maintain a harmonious cultural continuity: 'In an ideal state of society one might imagine the good New growing naturally out of the good Old, without the need for polemic and theory. This would be a society with a living tradition. In a sluggish society, as actual societies are, tradition is ever lapsing into superstition, and the violent stimulus of novelty is required.'[19] Here Eliot has hit upon a problem with his notion of tradition, which was to be clarified much later in *After Strange Gods*: namely, that tradition according to this scheme is *unconscious*. In the later work, Eliot was to introduce the concept of 'orthodoxy' as a conscious corrective, while preserving the cultural importance of habit and 'breeding'. As it appears in this earlier essay, the problem is still bound up with Eliot's own poetic ambitions: dare he 'disturb the universe' of European culture with his violent poetic novelties and endanger the very freedom from polemic and theory for which he values that culture? To this question – whether to let sleeping traditions lie – Eliot's answer here is only provisional, awaiting more sophisticated treatment in 'Tradition and the Individual Talent'. But what is glancingly revealed in the problem's terms, even as early as 1917, is Eliot's anxiety to preserve traditional culture from polemic and theory. The formulation recalls Arnold's praise of literary study for its avoidance of 'controversial reasonings', and points forward to Eliot's later statements on the place of the intellect in poetry.

One of the many intellectually acrobatic consequences of Eliot's timeless poetic order is that the present is able to alter the past, which is now simultaneous with it. Eliot often refers to his critical task as one of 're-ordering' the poetic tradition – which in most senses is a legitimate task of all literary historians and critics. But having, as it were, flattened history into a single ideal dimension, Eliot is able to do much more than just reassess various traditions: he can rearrange history, not according to any supposed trends of real historical development, but entirely according to aesthetic principles or tastes. It becomes possible, in other words, to create an inverted literary history in which the history, being secondary to the permanent and the poetic, is criticized and readjusted to accommodate it to the literary sensibility. Eliot's most famous exercise in this vein is the celebrated 'dissociation of sensibility' thesis, in which a difficulty in assessing the 'Metaphysical' poets calls forth a thumb-nail history of 'the English mind' and an implicit condemnation of the English Revolution:

If so shrewd and sensitive (though so limited) a critic as Johnson failed to define metaphysical poetry by its faults, it is worth while to inquire whether we may not have more success by adopting the opposite method: by assuming that the poets of the seventeenth century (up to the Revolution) were the direct and normal development of the precedent age; and, without prejudicing their case by the adjective 'metaphysical', consider whether their virtue was not something permanently valuable, which subsequently disappeared, but ought not to have disappeared. [SE, 285.]

Before the Revolution, poets 'possessed a mechanism of sensibility which could devour any kind of experience'. In Donne and Chapman, for example, 'there is a direct sensuous apprehension of thought, or a recreation of thought into feeling'. But in the seventeenth century 'a dissociation of sensibility set in, from which we have never recovered' [SE, 288]. Just how and why it 'set in' is left unexplained: Milton and Dryden are given the blame for aggravating it, but the hint contained in the reference to the Revolution was to be confirmed much later by Eliot when he finally rehabilitated Milton – the dissociation was not the fault of the poets, he revealed, but 'had something to do with the Civil War'.[20] It was not long before Eliot was publicly to identify his literary classicism with monarchism and Anglo-Catholicism in *For Lancelot Andrewes,* thus underlining the importance of the Civil War in his version of cultural and political history. But at this stage – 1921 – what is noteworthy is the nature of what could be called the prelapsarian sensibility in Eliot's scheme. Eliot explains the fall from 'direct sensuous apprehension of thought, or a recreation of thought into feeling' in terms of the difference between intellectual poets and reflective poets – a fine distinction whose status requires careful examination. As examples of the 'reflective' type he takes Tennyson and Browning:

Tennyson and Browning are poets, and they think; but they do not feel their thought as immediately as the odour of a rose. A thought to Donne was an experience; it modified his sensibility. When a poet's mind is perfectly equipped for its work, it is constantly amalgamating disparate experience; the ordinary man's experience is chaotic, irregular, fragmentary. The latter falls in love, or reads Spinoza, and these two experiences have nothing to do with each other, or with the noise of the typewriter or the smell of cooking; in the mind of the poet these experiences are always forming new wholes.[21]

In contrast with the merely reflective poet, it seems that the intellectual Donne is to be preferred for his capacity to synthesize experiences, or, in a characteristically physical metaphor, to *digest* them: Donne and his contemporaries 'could devour any kind of experience', while Tennyson and Browning merely performed a 'second order' function – they 'ruminated' [*SE*, 288]. This metaphor of involuntary incorporation seems to qualify strongly the conscious intellectual virtues which Eliot is supposedly expounding; and a further qualification comes in the passage immediately following, where philosophy and intellect are relegated to the status of convenient raw materials for poetry:

The possible interests of a poet are unlimited; the more intelligent he is the better; the more intelligent he is the more likely that he will have interests: our only condition is that he turn them into poetry, and not merely meditate on them poetically. A philosophical theory which has entered into poetry is established, for its truth or falsity in one sense ceases to matter, and its truth in another sense is proved. . . .

It is not a permanent necessity that poets should be interested in philosophy, or in any other subject. We can only say that it appears likely that poets in our civilization, as it exists at present, must be *difficult*. Our civilization comprehends great variety and complexity, and this variety and complexity, playing upon a refined sensibility, must produce various and complex results. The poet must become more and more comprehensive, more allusive, more indirect, in order to force, to dislocate if necessary, language into his meaning. [*SE*, 288-9.]

Plainly, Eliot emerges in this passage more than in any other as the author of *The Waste Land*, concerned to justify his strained and allusive poetic method. The argument is as adroit as the poem itself: seeking to refute the standard objections to 'metaphysical' poetry (the alleged 'quaintness' of its interests and philosophy), Eliot abandons the philosophy, while defending its poetic effect of difficulty, 'simultaneously' slipping from the seventeenth century to the present day to find the appropriate justification in the complexity of modern civilization. Like Arnold in 'The Modern Element in Literature', he argues that since there is a much greater range of experiences to be devoured, poetic continuity may have to be sacrificed to comprehensiveness. What is not adequately accounted for, however, is the form that this comprehensiveness takes: difficulty and allusiveness are seen as inevitable consequences of the complexity of the material to be embraced ('comprehended') by the poet, but there seems to be no room for *comprehension* in the intellectual sense. The frequent charges of obscurantism which are levelled at 'difficult' poetry, whether metaphysical or Eliotic, are thus not squarely met but side-stepped in this seminal statement. While he appears initially to be defending intellectual poets against a preference for reflective poets, Eliot concludes simply by defending difficult poets, without explaining why the difficulty is unresolvable, let alone discussing the intellectual faculties which might seem the most obvious agents of resolution and clarity, be it only to find them wanting.

iii. *Matter for argument*

The unproclaimed theory which accounts for the kind of arguments to be found in Eliot's early criticism is a modified version of Arnold's slogan, 'poetry is the reality, philosophy the illusion'. Like Arnold, who objected to Romantic poets (himself included) merely 'talking aloud', Eliot admires and seeks a poetic objectivity, stability, and comprehensiveness undisturbed by intrusions of debatable personal opinion or philosophical contention. But, armed with the theories of Symbolism and Imagism, and with the philosophy of F. H. Bradley, he has an alternative model to offer, more credibly engaging than Arnold's vague Hellenist nostalgia for serenity and high seriousness: the precise, palpable poetic image, to be taken as the product of a sensibility in which intellect and feeling are fused in the Bradleian unity of 'immediate experience'. Firmer and more tangible than Arnold's 'touch-stones', Eliot's recommended poetic virtues have the advantage of being associated insistently with an ideal sensibility which incorporates or assumes intellectual qualities, rather than sentimentally substituting for them as Arnold had done. Yet the result — a systematic exclusion of philosophy or 'belief' from serious consideration in the production and judgement of poetry — is not radically dissimilar.

The essay in which Eliot attempts to present these poetic virtues most carefully, in his tentative definition of 'wit', reveals this pressure towards the exclusion of 'belief' causing him considerable difficulties. For the protagonist of the piece is Andrew Marvell, Puritan and Revolutionist. Eliot has to go to great lengths of special pleading to abstract Marvell from this compromising political and ideological position:

The persons who opposed Charles I and the persons who supported the Commonwealth were not all of the flock of Zeal-of-the-land Busy or the United Grand Junction Ebenezer Temperance Association. Many of them were gentlemen of the time who merely believed, with considerable show of reason, that government by a Parliament of gentlemen was better than government by a Stuart; though they were, to that extent, Liberal Practitioners, they could hardly foresee the tea-meeting and the Dissidence of Dissent. [SE, 294.]

Rescued by Eliot from this minefield of Arnoldian caricatures, Marvell is 'but a lukewarm partisan' and 'more a man of the century than a Puritan'. This is awkward enough for the credibility of Eliot's historical scheme of 'dissociation'.[22] What is more significant is that the non-partisanship here attributed to Marvell is carried over into the definition of poetic 'wit' itself. The description of wit in terms of toughness and impersonality is, after *The Sacred Wood*, hardly unfamiliar; but a new note seems to be registered when Eliot disentangles wit from erudition and cynicism: 'It is confused with erudition, because it belongs to an educated mind, rich in generations of experience; and it is confused with cynicism because it implies a constant inspection and criticism of experience. It involves, probably, a recognition, implicit in the expression of every experience, of other kinds of experience which are possible ...' [SE, 303].

Set against Eliot's later (or even earlier) dogmatisms, this is a formulation distinctly open to 'liberal' interpretation — seeming to locate major poetic value in the avoidance of one-sided attitudes or partisan statements. Along with the rest of the essay, it represents an uncharacteristic concession to the orderly, gentlemanly side of English Protestantism, significantly stressing the 'internal equilibrium' [SE, 304] embodied in poetic wit. This quality of equilibrium takes us back to Arnold's constant concern with balance or poise, which is both a poetic goal and an accomplishment of socially liberalizing scope. The poetic ideal which it represents can be traced still further back, to the famous definition of the imagination in the fourteenth chapter of Coleridge's *Biographia Literaria,* which Eliot quotes to support his view:

'This power . . . reveals itself in the balance or reconcilement of opposite or discordant qualities: of sameness, with difference; of the general, with the concrete; the idea with the image; the individual with the representative; the sense of novelty and freshness with old and familiar objects; a more than usual state of emotion with more than usual order; judgement ever awake and steady self-possession with enthusiasm and feeling profound or vehement . . .' [*SE*, 298].

Far from being a peculiarly 'classicist' innovation Eliot's poetic ideal of wit, of the fusion of intellect and feeling in concrete images, has a respectable Romantic, even humanist, ancestry, as C. K. Stead and Frank Kermode have shown.[23] What is innovatory in Eliot's critical stance is the determination with which he seeks to displace in poetry what Wordsworth called the 'meddling intellect', by concentrating so exclusively upon this 'undissociated' quality of reconcilement in metaphor and image. It is significant that Eliot omits most of the first part of Coleridge's famous sentence, which actually reads: 'This power, first put in action by the will and understanding, and retained under their irremissive, though gentle and unnoticed, controul (*laxis effurtur hebenis*) reveals itself in the balance . . .'.[24] This omission is a suppression of the governing conscious element in poetic composition, or, in most of Eliot's formulations, a relegation of this element to the status of a secondary 'critical' labour of revision. It is this battle on behalf of the image against 'the undigested "idea", or philosophy' [*SW*, 67] which provides the constant theme of *The Sacred Wood* — a theme often understood too narrowly as one of personality and emotion. One of the clearest statements in the volume on this subject is this observation on Goethe's *Faust*: 'Goethe's demon inevitably sends us back to Goethe. He embodies a philosophy. A creation of art should not do that: he should *replace* the philosophy. Goethe has not, that is to say, sacrificed or consecrated his thought to make the drama; the drama is still a means.'[25] This statement, clearly implying that drama is an end in itself and that it should not even embody a philosophy, let alone express one, is certainly one of Eliot's most extreme. The essay in which it appears — 'The Possibility of a Poetic Drama' — is a sustained and provocative attack upon 'mixed art' on the stage, and in particular upon the Shavian 'theatre of ideas', to which he prefers mute theatre or the music hall. Eliot's theory is that the combination of ideas and art can lead only to mutual contamination unless ideas are so presented as to render them unrecognizable:

. . . the moment an idea has been transferred from its pure state in order that it may become comprehensible to the inferior intelligence it has lost contact with art. It can remain pure only by being stated simply in the form of general truth, or by being transmuted, as the attitude of Flaubert toward the small bourgeois is transformed in *Education Sentimentale*. It has there become so identified with the reality that you can no longer say what the idea is. [*SW*, 68.]

When trying to confirm the stature of a poet who is quite obviously committed to a certain philosophy, Eliot is forced to retreat from the inflexible positions expressed in this essay: Dante is the example which presses him repeatedly to modify his position on this question. His first attempt, in *The Sacred Wood*, reintroduces Arnold's division of labour between poet and philosopher, and gives a very useful precision of his notion of the 'transmutation' of ideas:

Without doubt, the effort of the philosopher proper, the man who is trying to deal with ideas in themselves, and the effort of the poet, who may be trying to *realize* ideas, cannot be carried on at the same time. But this is not to deny that poetry can be in some sense philosophic. The poet can deal with philosophic ideas, not as matter for argument, but as matter for inspection. The original form of a philosophy cannot be poetic. But poetry can be penetrated by a philosophic idea, it can deal with this idea when it has reached the point of immediate acceptance, when it has become almost a physical modification. [*SW*, 162-3.]

The distinction introduced here is noteworthy as a more considered statement of Eliot's objections to philosophical intrusions in poetry. What he is unwilling to accept is now not philosophy as such, but philosophy or ideas *as matter for argument*. If an idea is so 'digested', so stabilized into palpable presentation that it can hardly be identified, let alone disputed, it is acceptable; it conforms to the ideal fusion of intellect and feeling: 'We are not here studying the philosophy, we *see* it, as part of the ordered world.' [*SW*, 170.] For this immediacy of philosophy Eliot now introduces a fundamental precondition, implied in the passage just quoted and brought out clearly in the same essay where he justifies his estimate of Lucretius's superiority to Parmenides or Empedocles: 'He endeavours to expound a philosophical system, but with a different motive from Parmenides or Empedocles, for this system is already in existence.' [*SW*, 161.] This clarifies what was said about the work of the philosopher and that of the poet being carried on at the same time.

The poet in Eliot's view has to deal with ideas, if at all, when these ideas take their place in a more or less *finished* system, when they are absorbed, filtered unconsciously into a society's habits of perception. The negative example which is contrasted with Dante — that of William Blake — illustrates for Eliot the damaging crankiness of a poet attempting to philosophize on his own account:

His philosophy . . . was his own. And accordingly he was inclined to attach more importance to it than an artist should . . . The borrowed philosophy of Dante and Lucretius is perhaps not so interesting, but it injures their form less. Blake did not have that more Mediterranean gift of form which knows how to borrow as Dante borrowed his theory of the soul; he must needs create a philosophy as well as a poetry. [*SW*, 155-6.]

Blake's fundamental fault, for Eliot, was that he 'became too much occupied with ideas' [*SW*, 155.] 'What his genius required, and what it sadly lacked, was a framework of accepted and traditional ideas which would have prevented him from indulging in a philosophy of his own, and concentrated his attention upon the problems of the poet.' [*SW*, 157-8.] At first glance, it may appear that Eliot is using the same reproach that Arnold had levelled against the Romantics: that they 'did not know enough'. But further attention reveals that, as with Arnold, the accusation is an inverted censure for knowing too much. If the case of Blake is pressed just a step further than Eliot takes it, the question which arises is: if Blake had known more of traditional Latin culture (and he was not quite the Robinson Crusoe which Eliot paints him) would that knowledge have 'prevented him from indulging' in rebellion against that culture? There is at least enough doubt attached to this supplementary question to indicate that this was not the manner in which Eliot was really posing the problem. For him it was not so much a question of availability or choice of traditions, as one of prevention of poetic thinking which abandons the 'accepted and traditional ideas'; a preference for harmonious ignorance rather than nonconformist knowledge.

Eliot's position is a difficult one. He is not asking that all philosophy be banished from poetry — which could be the more or less consistently held position of an advocate of 'art for art's sake'. His argument is rather that only those philosophies which are traditional and accepted can be put to use by the poet.[26] At this stage, the question of the truth of a poet's philosophy is, for Eliot, entirely beside the point. Only the effect of ideas upon the poetic form is of any relevance to the problems of the poet. Eliot seems to be at a delicate transitional stage between a Symbolist purism of artistic autonomy and his later convictions of religious orthodoxy; a stage at which the stability of religious tradition is recognized and admired but as yet only from the irreverently aesthetic angle. Eliot is still, in other words, well within the Arnoldian framework of cultural criticism. The Arnold who had once described Catholicism as 'that great impersonal artist' [CPW viii. 350] is never very far from Eliot's mind in his early criticism, as this consideration of the background to Blake's case testifies:

We may speculate, for amusement, whether it would not have been beneficial to the north of Europe generally, and to Britain in particular, to have had a more continuous religious history. The local divinities of Italy were not wholly exterminated by Christianity, and they were not reduced to the dwarfish fate which fell upon our trolls and pixies. The latter, with the major Saxon deities, were perhaps no great loss in themselves, but they left an empty place; and perhaps our mythology was further impoverished by the divorce from Rome. Milton's celestial and infernal regions are large but insufficiently furnished apartments filled by heavy conversation; and one remarks about the Puritan mythology an historical thinness. And about Blake's supernatural territories, as about the supposed ideas that dwell there, we cannot help

commenting on a certain meanness of culture. They illustrate the crankiness, the eccentricity, which frequently affects writers outside of the
Latin traditions, and which such a critic as Arnold should certainly
have rebuked. [*SW*, 157.]

The comparison, 'for amusement', of religious traditions solely
in terms of the impressiveness or style of their mythologies,[27]
and the nostalgia for pre-Reformation, pre-revolutionary
stability are indeed Arnold's. For both critics the enemy is
Dissent, both with a large and with a small 'd' – not only in
religion, that is, but in philosophy and in literary criticism
itself, since criticism above all has to observe an 'objective'
order: 'we perceive that criticism, far from being a simple and
orderly field of beneficent activity, from which imposters
can be readily ejected, is no better than a Sunday park of
contending and contentious orators, who have not even arrived
at the articulation of their differences. . . . We are tempted to
expel the lot.' [*SE*, 25.]

This distaste for controversy and partisanship, with its
overtones of political despotism, is one of the most prominent
common characteristics of the two critics. Eliot's distinction
was to have reasserted the need for a cultural 'orthodoxy' at
a moment of acute social disintegration; to have re-enacted
Arnold's labours while taking pains for a short while at least
to avoid the temptations which had undermined Arnold's
'disinterested' stance. The few years in which Eliot was able
to discipline his social, political, and religious convictions,
'digesting' or channelling them into a strictly defined literary-
critical discourse, are generally acknowledged as the period
of his most influential critical writing. This effort of concentration owed much to the compression necessary for the
book review (in which form most of his early criticism
appeared) and to the fact that Eliot was a practising poet,
concerned to direct a whole range of wider problems towards
the 'problems of the poet'. Unlike many of the established
critics of his day, Eliot was not (apart from a few early years
of extension lecturing) a teacher; in fact much that is new in
his early criticism is a reaction against the personal and
historical 'heresies' of the extension-lecture tradition.[28] Yet
the timely critical strength involved in this distance from the
English academic establishment and its routines was to

become a weakness once Eliot's critical 'order' began to come apart under the pressure of his political and religious beliefs. Dismissing Arnold's educational progressivism, Eliot failed to discern, as Arnold had discerned, the importance of the educational system as an agency of cultural continuity. As a result of this failure, he proved incapable of carrying through any sustained cultural project of wider scope than the tiny readership of the *Criterion*. Instead it was to be the university English studies movement which showed itself able to expand upon Eliot's early promise, appropriating his critical innovations — often against his will — and using them for its own purposes.

NOTES

1. *For Lancelot Andrewes*, ix.
2. *Syllabus* . . . rpt. in Ronald Schuchard, 'T. S. Eliot as an Extension Lecturer 1916-1919' *Review of English Studies*, xxv (1974), 165.
3. See Ronald Schuchard, 'Eliot and Hulme in 1916: Toward a Revaluation of Eliot's Critical and Spiritual Development', *PMLA*, lxxxviii (1973), 1083-94.
4. *Egoist*, v (1918), 1 ('In Memory of Henry James').
5. F. W. Bateson, 'The Poetry of Learning', in G. Martin (ed.), *Eliot in Perspective* (1970), 42.
6. *The Use of Poetry and the Use of Criticism*, 109.
7. Watson, *Literary Critics*, 177-8.
8. *Egoist*, v. 1 ('In Memory of Henry James').
9. Ibid., 2.
10. *Egoist*, v. 69 ('Observations' — signed 'T. S. Apteryx').
11. *Egoist*, iv (1917), 118 ('Reflections on Contemporary Poetry I').
12. *Egoist*, v. 1 ('In Memory of Henry James').
13. *Egoist*, v. 61 ('Professional, Or . . .' — signed 'Apteryx').
14. *Egoist*, v. 132 ('Studies in Contemporary Criticism II').
15. *SE*, 27. Cf. Arnold, *CPW* iii. 288-9 (Preface to *Essays in Criticism*, 1865) and the third section of Eliot's own 'East Coker'.
16. *SE*, 458. There is possibly some unconscious connection at the source of Eliot's revulsion, between the 'rapid-breeding' cinema and his image, mentioned earlier, of ideas as rabbits running wild.
17. *After Strange Gods* (1934), 15.
18. *SW*, 49-50. Watson remarks pointedly that 'it is an odd historical sense that denies chronology and conceives of the past both as "the timeless and the temporal"; it might have been franker, one begins to

feel as the essay proceeds, to call it "the anti-historical sense"' (Watson, *Literary Critics*, 172).

19. *To Criticize the Critic* (1965), 184.

20. *On Poetry and Poets* (1957), 153.

21. *SE*, 287. The formulation appears to owe a great deal to Herbert Grierson's Introduction to *Metaphysical Poems and Lyrics of the Seventeenth Century* (Oxford, 1921), of which Eliot's essay was a review. Grierson had written of George Herbert's 'imaginative apprehension of emotional identity in diverse experiences, which is the poet's counterpart to the scientific discovery of a common law controlling the most divergent phenomena' (pp. xliv–xlv) — and had cited this as an illustration of Coleridge's concept of the Imagination. Spinoza's appearance in this passage is somewhat odd, if it is recalled that Arnold's enthusiasm for this philosopher rested partly on the fact that 'the ordinary man' could *not* read him.

22. Eliot seems to have been the first critic seriously to challenge his own 'dissociation' thesis, or at least its chronology: as early as 1924 he was to write, 'Even the philosophical basis, the general attitude toward life of the Elizabethans, is one of anarchism, of dissolution, of decay . . . The Elizabethans are in fact a part of the movement of progress or deterioration which has culminated in Sir Arthur Pinero and in the present regiment of Europe.' [*SE*, 116–17.] The implication of this is to back-date the great cultural disaster from the seventeenth century to (at least) the Reformation. Again in 1929 Eliot wrote that 'the process of disintegration which for our generation culminates in that treaty [of Versailles] began soon after Dante's time' [*SE*, 240]. As Raymond Williams has remarked (*The Country and the City* (1973), 11–12), referring to F. R. Leavis's (and others') 'organic community', this kind of back-dating stops only at Eden.

23. C. K. Stead, *The New Poetic* (1964); Frank Kermode, *Romantic Image* (1957).

24. S. T. Coleridge, *Biographia Literaria*, ed. J. Shawcross (1907), ii. 12.

25. *SW*, 65–6. This, like Eliot's contrast between Dante and Blake, is a direct attack upon the 'criticism of life' position put forward by Arnold in *The Study of Celtic Literature*: 'Goethe's task was, — the inevitable task for the modern poet henceforth is . . . not to preach a sublime sermon on a given text like Dante, but to interpret human life afresh, and to supply a new spiritual basis to it.' [*CPW* iii. 381.] Eliot's objection to Goethe leads him to the defence of Dante and the reliance of poetry upon a 'given text' or belief.

26. Possibly the clearest statement of this doctrine was made by Eliot years later: 'I believe it to be a condition of success, that the view of life which [poets] attempt to express in poetry, should be one which is already accepted. I do not think you can make poetry out of ideas when they are too *original*, or too *new* . . . For the business of the poet is to express the culture in which he lives, and to which he belongs, not

to express aspirations towards one which is not yet incarnate.' *Adelphi*, xxi (1945), 154 ('The Social Function of Poetry').

27. Even when Eliot came to reject the substitution of art for religion, it seemed that his motives were not those of doctrinal principle, but of good taste. In his 'Dialogue on Dramatic Poetry', for example, the view that the Mass is the highest form of drama is rejected on purely literary-critical grounds as a *'confusion des genres'*. [*SE*, 248]. An interest in the fate of the pre-Christian European divinities had been an important component of the semi-religious aestheticism of Walter Pater.

28. Some of Eliot's earliest protests against the inadequacies of English criticism refer to his experience of extension lectures. In an article of 1918 he attacks the amateurism of Mrs Meynell by saying that her writing 'is what a University Extension audience would like; but it is not criticism' [*Egoist*, v. 61 ('Professional, Or . . .')]. Arthur Symons is berated in similar terms in 'The Perfect Critic'.

6. LITERARY-CRITICAL CONSEQUENCES OF THE PEACE: I. A. RICHARDS'S MENTAL LEAGUE OF NATIONS

Prohibit sharply the rehearsed response
And gradually correct the coward's stance.
W. H. Auden

At Cambridge, the generation of students returning from the war brought with them a suspicion of their elders which provided fertile ground for Eliot's new unsentimental attitude to literature. The disorganization caused by the war had meant, moreover, that in the long absences of Quiller-Couch, the new English School was discreetly shaped by a cabal of younger men dedicated to an experimental approach to literary studies; an approach befitting the clean sweep which post-war reconstruction seemed to require.[1] Basil Willey, who arrived at Cambridge with this generation, recalls this demand for poems fit for heroes:

Many of us were just back from the 1914–18 war, and we were unsatisfied with the woolly generalities and the vague mysticism of the accepted schools of criticism (Gosse, Symons, Saintsbury and the like). The English Tripos had only just been founded, and we wanted to feel that we were not merely going to tread the hackneyed roads leading from Aristotle to Croce, but were in at the start of an exciting new enterprise and about to break new ground. . . . Richards gave us then, as often since, the sense that a dawn was breaking in which it would be bliss to be alive, a new day in which knowledge of our own minds would give us enhanced control over ourselves and our destinies; and in which poetry, as the storehouse of recorded values, would be seen as the stronghold of the spirit of man in its struggle for the good life in the face of scientific aggression.[2]

I. A. Richards himself, over fifty years later, described the mood at Cambridge upon which such hopes were built, as he had encountered it when he started teaching there in 1919: 'That year was quite beyond anything you could imagine. It was World War I survivors come back to college. . . . There was an atmosphere, such a dream, such a hope. They were just too good to be true; it was a joy to deal with those people; those who got back to Cambridge from all that slaughter were back *for reasons*.'[3]

This enormous sense of expectation had been foreseen in the English School.[4] Of the 'new blood' brought in on the teaching side to meet the demand, Richards had the advantage of a training outside literary studies – primarily in moral philosophy. He arrived, therefore, with a habit of asking questions which the routine mysticism of established critics like Symons or Saintsbury had all but entirely neglected. As the only trained philosopher in the school, his assertions were peculiarly invulnerable to challenge, and in the eyes of a sceptical generation of students, his comparative rigour enabled him to emerge as a figure endowed, in Basil Willey's words, with 'the qualities of a religious leader, prophet or guru'.[5] According to William Empson, Richards's lectures were sometimes so oversubscribed that he had to hold them in the street, an event said to have been unprecedented since the Middle Ages.[6] Launched upon this tide of enthusiasm, Richards's literary-critical work – in particular *Principles of Literary Criticism* (1924), *Science and Poetry* (1926), and *Practical Criticism* (1929) – was to place its mark indelibly upon the study of literature in the English-speaking world. It was itself marked, however, by the conditions under which it was launched, and in ways whose importance for an understanding of his thought needs more detailed emphasis.

For Richards, one of the most alarming effects of the war had been the way in which it had revealed society's apparently limitless capacity for self-deception. The many ways in which experience can be misrecognized or distorted under the pressure of unacknowledged motives, interests, and inhibitions (which he was to catalogue in detail in connection with the misreading of poems) became a preoccupation, whose moral basis is to be found in his attitude to the war. His search for methods of undermining 'stock responses' followed from his awareness of the fearful dishonesty of wartime propaganda. 'In war-time words become a normal part of the mechanism of deceit',[7] wrote Richards and C. K. Ogden in *The Meaning of Meaning* (1923), a work actually conceived on Armistice Day itself; and to them, the deceit appeared to be only in its early stages: 'The return of the exploiters of the verbal machine to their civil posts is a return in triumph, and its effects will be felt for many years in all countries where the power of the word among the masses remains paramount.'[8] *The Meaning of Meaning* is an elaborate attempt to limit the power of such 'word magic', to separate the scientific and emotive uses of language whose illegitimate mixture in wartime propaganda appeared to have stirred such mass irrationality. Richards was committed to salvaging from the experience of the war some lessons which could help avoid future catastrophes. The urgency of his desire that this experience should not go to waste is felt fully in his discussion of that variety of 'inappropriate response' known as sentimentality:

The best example is the pathetic and terrible change that can too often be observed in the sentiments entertained towards the War by men who suffered from it and hated it to the extremist degree while it was raging. After only ten years they sometimes seem to feel that after all it was 'not so bad', and a Brigadier-General recently told a gathering of Comrades of the Great War that they 'must agree that it was the happiest time of their lives'. [*PC*, 260-1.]

Richards only rarely makes such direct references to the war, but the reserves of pity and anger mobilized on such occasions contrast with the blandly good-humoured tone of the rest of his writings sharply enough to suggest how important this catastrophe was to his sense of critical purpose.

i. *Chaos and its inhabitants*

One of Richards's most widely quoted remarks upon an individual poem — that Eliot's *The Waste Land* expresses 'the plight of a whole generation' [*PLC*, 235] — affirms the constant sense of timeliness in his writings; the conviction that the war had placed momentous responsibilities and choices before the survivors. It is characteristically in the vocabulary of the war and its settlement (or postponement) that Richards most clearly summarizes his view of this turning-point in history and of the problems which it poses:

In the past, Tradition, a kind of Treaty of Versailles assigning frontiers and spheres of influence to the different interests, and based chiefly upon conquest, ordered our lives in a moderately satisfactory manner. But Tradition is weakening. Moral authorities are not as well backed by belief as they were; their sanctions are declining in force. We are in need of something to take the place of the old order. Not in need of a new balance of power, a new arrangement of conquests, but of a League of Nations for the moral ordering of the impulses; a new order based on conciliation, not on attempted suppression.[9]

It is very probable that the Hindenburg line to which the defence of our traditions retired as a result of the onslaughts of the last century will be blown up in the near future. If this should happen a mental chaos such as man has never experienced may be expected. We shall then be thrown back, as Matthew Arnold foresaw, upon poetry. It is capable of saving us; it is a perfectly possible means of overcoming chaos. [*SAP*, 82-3.]

What is concentrated in these statements, by means of the striking political and military metaphors, is the core of Richards's critical concern: the safeguarding of cultural order against a threatened chaos through the conciliatory agency of poetry. This concern, evidently that of Arnold himself, of the Newbolt Report,[10] and (with certain important qualifications) of T. S. Eliot, is underlined throughout Richards's major writings of the 1920s by a repeated assertion of the dangers and threats lurking in the state of contemporary culture. *Science and Poetry*, with its warnings of 'chaos', is the most consistently alarmist statement of this view, and the most far-fetched in the hopes it places in poetry as a remedy; but the more carefully argued, and subsequently more influential works, *Principles of Literary Criticism* and *Practical Criticism*, derive their sense of purpose and urgency from the same sources of social and cultural anxiety.

Principles of Literary Criticism, which Richards introduced in his 1928 Preface as 'a loom upon which it is supposed to re-weave some ravelled parts of our civilisation', addresses itself extensively to the causes of the impending threat and to the major role of the literary critic in averting it. Richards argues that in the present state of culture, a growing mass of people threatens to swamp the tiny cultured minority:

With the increase in population the problem presented by the gulf between what is preferred by the majority and what is accepted as excellent by the most qualified opinion has become infinitely more serious and appears likely to become threatening in the near future. For many reasons standards are much more in need of defence than they used to be. It is perhaps premature to envisage a collapse of values, a transvaluation by which popular taste replaces trained discrimination. Yet commercialism has done stranger things: we have not yet fathomed the more sinister potentialities of the cinema and the loudspeaker . . .[11]

What is even worse for Richards than this minefield of dangers is that those who represent qualified opinion (and who seem unfortunately not to breed as rapidly as the rest) are unprepared to face the threat. They lack a 'defensible position', a 'stronghold', and 'weapons with which to repel and overthrow misconceptions' [PLC, 25]. Their tastes, in the absence of any clear justification for them, appear as signs of a snobbishness which only encourages the suspicion and hostility of the masses.

To bridge the gulf, to bring the level of popular appreciation nearer to the consensus of best qualified opinion, and to defend this opinion against damaging attacks . . . a much clearer account than has yet been produced, of why this opinion is right, is essential. These attacks are dangerous, because they appeal to a natural instinct, hatred of 'superior persons'. The expert in matters of taste is in an awkward position when he differs from the majority. He is forced to say in effect 'I am better than you. My taste is more refined, my nature more cultured, you will do well to become more like me than you are.' It is not his fault that he has to be so arrogant. He may, and usually does, disguise the fact as far as possible, but his claim to be heard as an expert depends upon the truth of these assumptions. He ought then to be ready with reasons of a clear and convincing kind as to why his preferences are worth attention, and until these reasons are forthcoming, the accusations that he is a charlatan and a prig are embarrassing. . . .

To habilitate the critic, to defend accepted standards . . . to narrow the interval between these standards and popular taste, to protect the arts against the crude moralities of Puritans and perverts, a general theory of value . . . must be provided. There is no alternative open.[12]

It was within this framework of defending accepted minority tastes against a growing army of enemies that Richards's theory of value was formulated. Of those enemies, we have already seen mentioned the cinema, the democratic multitude suspicious of 'superior persons', the collapse of traditional beliefs, moral Puritans, and unspecified 'perverts'. Elsewhere, Richards expands upon his idea of the cultural threat, seeing it as the result of a too great accumulation of new ideas over the years, combined with a sudden spreading of those ideas which breaks down small enclosed communities. Scraps of different cultures are thereby flung together, numbing our understanding of the whole. 'We defend ourselves from the chaos that threatens us by stereotyping and standardising both our utterances and our interpretations. And this threat, it must be insisted, can only grow greater as world communications, through the wireless and otherwise, improve.' [*PC*, 340.] Richards repeatedly identifies modern communications media — the radio and the cinema — with the coming cultural chaos, blaming them for a rapid and confusing dissemination of new ideas:

It is arguable that mechanical inventions, with their social effects, and a too sudden diffusion of indigestible ideas, are disturbing throughout the world the whole order of human mentality, that our minds are, as it were, becoming of an inferior shape — thin, brittle and patchy, rather than controllable and coherent. It is possible that the burden of information and consciousness that a growing mind has now to carry may be too much for its natural strength. If it is not too much already, it may soon become so, for the situation is likely to grow worse before it is better. [*PC*, 320.]

For Richards, the source of the modern predicament is (as it was for Arnold) the too sudden diffusion of new ideas among large masses of people who are not mentally stable enough to bear their weight. Not only do the masses suffer from confusion, but the ideas themselves become coarsened from widespread handling. Richards argues that there is an inevitable 'loss in transmission' when ideas are circulated widely; that 'any very widespread diffusion of ideas and responses tends towards standardisation, towards a levelling down' [*PC*, 248] . Accordingly, the great majority of people are assumed to be caught up in a fantasy world of 'stock responses', stereotyped ideas, and emotional attitudes fixed for them by the mass media. Any suggestion that there may be a corresponding 'gain' in transmission, whereby new ideas may be enriched or adapted by the multitude, is excluded in

Richards's account. His comments on advertising imply a complete identification of the advertisement and the intended audience, as if the mass audience were accountable for the advertisement's faults.

Richards goes so far as to assert that advertisements 'undoubtedly represent the literary ideals present and future of the people to whom they are addressed' [*PLC*, 159–60]. One of the very few occasions on which he holds out the possibility of criticism becoming a science is made in this connection: 'Criticism will justify itself as an applied science when it is able to indicate how an advertisement may be profitable without necessarily being crass.' [*PLC*, 160.] Such a statement casts a good deal of light on the 'practical' intention of Richards's literary-critical project. Breaking boldly away from the 'ivory tower' of detached literary gossip and commentary, his world-reforming aspirations seem to dwindle into the tasteful rehabilitation of advertising agencies. Without any considered theory of the society he is trying to reform (apart from a very impressionistic amateur-sociological notion of 'mass culture'), Richards allows himself simply to assume that the crassness of advertisements is a problem detachable from their profitability. Elsewhere, he sees the culprit as 'commercialism', but here it becomes clear that commercialism's faults are reducible to the bad taste of its consumers. The practical aim of Richards's cultural mission is therefore to correct the taste of the masses while leaving commercial institutions unchallenged.

Richards's view of culture clearly implies that it is from the large, infinitely suggestible, and emotionally immature mass of people that the greatest threat is likely to emerge. In his discussion of stock responses, he implies that the mass of the population would be 'easy to control by suggestion' [*PC*, 314], should a minority decide what kinds of suggestions to disseminate. Large meetings and processions (common enough occurrences in the troubled economic climate of the 1920s) always involve 'emotional orgies' of mass suggestion and sentimentality which we only recognize when we 'escape from the crowd' [*PC*, 257]. While the great poets attain a supreme level of self-control and self-understanding, the supposedly 'educated' classes of society deceive themselves that they possess these qualities; 'As to the less 'well-educated',' Richards remarks, 'they inhabit chaos.' [*PC*, 325.] The threatened chaos to which Richards repeatedly refers as the occasion for his critical campaign is indeed closely identified in his writings with this class of people, as it had been by Eliot and Arnold before him.

Like Arnold, who saw his 'Populace' as a kind of collective personality, a grasping intolerant self pitted against the 'best self' of the State, Richards approaches the culture of the majority as a problem of individual psychology (indeed, psychiatry), associating widespread 'stock responses' with emotional immaturity and inhibition. His entire critical project is presented as a large-scale exercise in mental health. Richards asserts that 'the critic is as closely occupied with the health of the mind as any doctor with the health of the body' [*PLC*, 25]. His generation suffers from various kinds of nervous stress resulting from the vain attampt to patch up the old beliefs with science alone [*PLC*, 221]. The social and cultural chaos which he sees looming is, in his view, a product of too many chaotically disorganized minds, not only among the majority who 'inhabit chaos', but even in the supposedly educated classes.[13] In addressing this problem, Richards seeks to use the lessons of neurology ('the only good legacy left by the War' [*PLC*, 63]), and of psychology. He resolves to approach the critical samples collected in *Practical Criticism* with 'a few touches of the clinical manner' [*PC*, 7], comparing the writings of his students with the ravings of lunatics.

Richards's argument — the core of his value-theory — is that while the ordinary mind suppresses nine-tenths of its 'impulses' (or experiences)[14] because it is unable to digest them without confusion, the 'superior man' [*PC*, 287] — and this usually means the artist — can achieve a mental poise or equilibrium whereby far more 'impulses' can be accommodated into an ordered, self-complete whole. Referring back to Coleridge's description of the reconciling properties of the imagination,[15] as Eliot had done in his discussion of the undissociated sensibility, Richards stresses the wholeness of the mind, both of the author and of the reader, as the secret of poetry and of the arts in general. But this theory of balanced impulses reaches far beyond mere explanation of the workings of poetic imagination. Richards asserts that such a 'finer organization [of the mind] is the most successful way of relieving strain, a fact of relevance in the theory of evolution. The new response will be more advantageous than the old, more successful in satisfying varied appetencies.' [PLC, 153.] Taking into account Richards's statements that poetry is capable of saving us from impending biological calamities, it can be inferred from this remark that the poet (the 'superior man') is in the vanguard of humanity's very biological evolution, and that the wholeness of mind which we achieve from reading poetry is somewhere at the centre of any resolution of the general social and cultural crisis besetting the species.

ii. 'The most insidious perversion'

Essential to his view of poetry as an agent of mental health is Richards's insistence upon the *normality* of the artist. Like Wordsworth, Richards sees poets and artists as differing from other people by degree rather than by kind: they assimilate more impulses, not impulses of a different order. If poets are in some way more advanced upon the evolutionary scale, it is not as mutants but as normal specimens who have adjusted themselves to incorporate experiences readily available to all other people. It is the 'Puritans and perverts' rather than the artists who deviate from the balanced human norm, by misshaping their minds into narrow and exclusive channels, squandering their full capacities by such exclusion. To the degree to which we all suppress 'impulses' and surrender to mental confusion and one-sidedness we are all less normal than the artist. To the degree to which we fail to appreciate great poetry we are, in Richards's view, abnormal and wasteful in the rest of our lives. 'The basis of morality, as Shelley insisted, is laid not by preachers but by poets. Bad taste and crude responses are not mere flaws in an otherwise admirable person. They are actually a root evil from which other defects follow.' [PLC, 47.]

The 'crude response' registered by failure to appreciate great poetry brands us, according to Richards, as psychologically immature, inhibited, and incomplete. He cites the 'erratic' reader 'familiar to every teacher concerned with poetry', whose main traits of character are 'obstinacy and conceit' caused by 'some disorder of the self-regarding sentiment — a belated Narcissism, perhaps' [PC, 250-1]. It seems that the achievement of a unified sensibility, an ordered mind free from wasteful and disordering inhibition, is threatened by sinister obstacles set in the way by the perversities of the ordinary human mind. Consequently, an approach to poetry which is satisfied with simple 'immersion' in the verse will fail to achieve this supremely valuable state. Just as Arnold had claimed in his essay on academies that a high standard of taste 'has many enemies in human nature' [CPW iii. 235], Richards elaborates: 'The alluring solicitancy of the bad, the secret repugnancy of the good are too strong for us in most reading of poetry. Only by penetrating far more whole-mindedly into poetry than we usually attempt, and by collecting all our energies in our choice, can we overcome these treacheries within us. That is why good reading, in the end, is the whole secret of "good judgement".' [PC, 305.]

In his repeated implication that the reading of good poetry — which embodies the finest organization of impulses and hence the greatest value in life — is essential to the sanity of life in general, Richards does not mean that we can attain the artist's supreme normality without intense effort. If the fine mental organization embodied (as it is nowhere else) in poetry is the real basis of morality, then the same self-discipline required by other moral codes becomes essential in our approach to poetry. There are internal treacheries to overcome, of which the most formidable in Richards's view is the mental affliction whereby we attempt to tie our beliefs and attitudes to established facts:

The central experience of Tragedy and its chief value is an attitude indispensable for a fully developed life. But in the reading of *King Lear* what facts verifiable by science, or accepted and believed in as we accept and believe in ascertained facts, are relevant? None whatever. Still more clearly in the experiences of some music, of some architecture and of some abstract design, attitudes are evoked and developed which are unquestionably independent of all beliefs as to fact, and these are exceptional only in being protected by accident from the

most insidious perversion to which the mind is liable. For the intermingling of knowledge and belief is indeed a perversion, through which
both activities suffer degradation. [*PLC*, 223]

From Richards, such strong language indicates a concern at
the very heart of his argument for poetry as a saviour. If the
intermingling of knowledge and belief is indeed 'the most
insidious perversion to which the mind is liable', then it is
a major obstacle to mental health and consequently a major
factor in the threat of social chaos.

He asserts flatly that there 'are two totally distinct uses of
language' [*PLC*, 206]. One refers to external facts and is the
basis for science. The other expresses feelings and attitudes
and is the basis for poetry and religion. These are the referential
and the emotive uses of language, which produce, respectively,
factual statements and 'pseudo-statements' (or questions
beginning 'How?' answerable by statements, and questions
beginning 'What?' or 'Why?' which are simply requests for
emotional satisfaction, not real questions). According to
Richards, the emotive function combines impulses in a
manner entirely free from logical relations. The cardinal
perversion of the modern mind is to confound these two uses
of language — in particular the attempt to seek a factual basis
for our attitudes, which is an entirely unnecessary quest since
a pseudo-statement 'is "true" if it suits and serves some
attitude or links together attitudes which on other grounds
are desirable'.[16] In other words attitudes, to the extent that
they are part of the mind's continuous self-adjustment, are
their own self-justification; and the search for factual truth
to guarantee them is misguided and self-destructive. Richards
regards such mixtures as explosive, and as a particularly
dangerous feature of the modern sensibility which requires
the opening of an interim critical period before a later
synthesis can be achieved:

Mixed modes of writing which enlist the reader's feeling as well as his
thinking are becoming dangerous to the modern consciousness with its
increasing awareness of the distinction. Thought and feeling are able to
mislead one another at present in ways which were hardly possible six
centuries ago. We need a spell of purer science and purer poetry before
the two can again be mixed, if indeed this will ever become once more
desirable. [*PLC*, 1928 Preface.]

Just why Richards devotes so much attention to this distinction may become clearer if his concept of the 'attitude' is examined more closely. In his theory the term has a specific meaning which is defined above all by its contrast with action. Attitudes are those preliminary organizations which we adopt in adjusting ourselves in readiness for possible action. Richards sees it as very important that these adjustments can be made without any overt action necessarily ensuing – only 'incipient' or 'imaginal' action. 'But imaginal action and incipient action which does not go so far as actual muscular movement are more important than overt action in the well-developed human being. Indeed the difference between the intelligent or refined, and the stupid or crass person is a difference in the extent to which overt action can be replaced by incipient and imaginal action.' [*PLC*, 85.] His distinction between scientific beliefs and 'emotive' belief hinges upon this very difference. He defines scientific belief as 'readiness to act as though the reference symbolized by the proposition which is believed were true' [*PLC*, 219] in all cases. Emotive beliefs, and in particular the statements of poetry or drama, however, are entirely conditional, taking the form 'given this, then that would follow'; and readiness to act upon these beliefs is therefore restricted to the 'imaginal' internal adjustments which Richard calls attitudes. The function of poetry is to exercise or give vent to these attitudes without spilling over into action. With this understood it should be possible to see why the confusion of emotive and referential functions is regarded as such a major 'perversion': it leads to, or at least justifies, the taking of action on behalf of scientifically unproven beliefs. As Richards stresses in *Science and Poetry* (it is significantly the only italicized sentence in the whole book): '*In a fully developed man a state of readiness for action will take the place of action when the full appropriate situation for action is not present.*' [*SAP*, 20.] Where the refined person can respond to events with a peaceful self-adjustment, the crass person is always – to use Arnold's phrase – rushing out into the streets with half-formed ideas or 'pseudo-statements', engaging in premature and chaotic activity.

Richards is so convinced of the socially dangerous

consequences of this maladjustment that he claims, of the distinction between scientific and emotive language, that 'No revolution in human affairs would be greater than that which a widespread observance of this distinction would bring about.' [*PLC*, 217.] In attempting to initiate this revolution, he adopts a strategy almost identical to Arnold's: the disengagement of ideologies or attitudes from questions of truth or fact, with the aim of preserving them in their own privileged sphere from which they may take command of life unassailed by scientific scrutiny. He observes that under the impact first of natural science and now of social science, important stabilizing 'pseudo-statements' about God and human nature are no longer believed to be true,

. . . and the knowledge which has killed them is not of a kind upon which an equally fine organisation of the mind can be based.

This is the contemporary situation. The remedy, since there is no prospect of our gaining adequate knowledge, and since indeed it is fairly clear that genuine knowledge cannot serve us here and can only increase our practical control of Nature, is to cut our pseudo-statements free from belief, and yet retain them, in this released state, as the main instruments by which we order our attitudes to one another and to the world. Not so desperate a remedy as may appear, for poetry conclusively shows that even the most important among our attitudes can be aroused and maintained without any belief entering in at all. . . . It is only when we introduce illicit beliefs into poetry that danger arises. To do so is from this point of view a profanation of poetry. [*SAP*, 60-1.]

The 'danger' once again invoked by Richards is that questions of truth and belief may contaminate the free-floating poetic sphere, producing a mixture of convictions and emotions which could prove socially explosive. Richards's anxieties (again, very Arnoldian ones) about the too sudden spread of new ideas among a populace unable to digest them harmoniously are closely related to his segregation of beliefs and emotions. He seeks to avoid a general hardening of attitudes by placing traditional social values outside the range of scientific disproof. For Richards the jettisoning of dogmatic ballast would greatly increase these values' chance of survival, since only flexible adjustment, rather than rigid entrenchment, can withstand the coming shocks:

The present lack of plasticity in such things [moral principles] involves a growing danger. Human conditions and possibilities have altered more in a hundred years than they had in the previous ten thousand, and the next fifty may overwhelm us, unless we can devise a more adaptable morality. The view that what we need in this tempestuous turmoil of change is a Rock to shelter under or to cling to, rather than an efficient aeroplane in which to ride it, is comprehensible but mistaken. [*PLC*, 43.]

The virtues of the 'aeroplane', as they appear in this theory, are those of the supple mind, 'the mind that can shift its viewpoint and still keep its orientation'. This mind, which can transform itself without confusion or strain, is identified by Richards as 'the mind of the future' [*PC*, 343]. For the development of such suppleness of mind, unencumbered by the rigidities of belief, familiarity with great poetry is essential, since it is poetry above all which in his view proves the possibility of useful attitudes free from all beliefs, or self-adjustments without overt activity. The equilibrium or 'synaesthesia' of impulses achieved through poetry lifts us beyond our personal beliefs or convictions and makes us, Richards claims in a doubly resonant phrase, 'impersonal or disinterested'.[17] Just as with Arnold, the flexible suspension of poetry enables us to change our ground without any turmoil of controversial reasonings. To a defender of threatened minority standards, such abilities could indeed appear capable of 'saving us'.

iii. *The most valuable people*

An important consequence of substituting poetry-appreciation for the old and disintegrating religious and moral beliefs is that it begins to take on, in the minds of its advocates at least, many of the attributes of what it is replacing. Notions of poetry as a saviour or as a form of religion were of course already current, as the Newbolt Report and Sampson's *English for the English* show. In Richards's case, hints of a similar attitude may be found towards the end of *Practical Criticism*, where he discusses the benefits of poetry in terms of the Confucian concepts of sincerity and self-completeness, positing the existence within us all of 'a tendency towards increased order' [*PC*, 285] — which in this context strongly resembles Arnold's definition of God as a tendency towards righteousness. Later in this discussion Richards recommends that we might use 'a technique or ritual for heightening sincerity' [PC, 290] whenever we are unsure whether our response to a poem is genuine, a ritual involving deep contemplation of certain aspects of the human condition. He even admitted after many years that *Principles of Literary Criticism* was itself a 'sermon'.[18] But whether or not his concept of poetry's importance is decked out with such devotional vocabulary, it is clear that for him poetry takes on responsibilities of a scale formerly entrusted only to religions — from the achievement of individual peace of mind to the safeguarding of civilizations. Richards's most sustained assertion of the importance of the arts, in the fourth chapter of *Principles of Literary Criticism,* states the argument thus:

The arts are our storehouse of recorded values. They spring from and perpetuate hours in the lives of exceptional people, when their control and command of experience is at its highest, hours when the varying possibilities of existence are most clearly seen and the different activities which may arise are most exquisitely reconciled, hours when habitual narrowness of interests or confused bewilderment are replaced by an intricately wrought composure. . . . They record the most important judgements we possess as to the values of experience . . . without the assistance of the arts we could compare very few of our experiences. . . . In the arts we find the record in the only form in which these things can be recorded of the experiences which have seemed worth having to the most sensitive and discriminating persons.[19]

The experiences offered by the arts are unobtainable by other means; where other experience is 'incomplete', art completes it. And the most complete people are 'precisely the people who most value these experiences' [PLC, 184]. It is through the arts above all that 'men may most co-operate' [PLC, 102] and the mind organize itself. The arts are 'the supreme form of the communicative activity' [PLC, 17] and are so central to all human values that the questions 'What is good?' and 'What are the arts?' cannot, in Richards's view, be answered separately [PLC, 27]. Given the importance of language, poetry occupies a special place above the other arts: it is 'the most important repository of our standards' [PC,

49], compared with which 'every other state of mind is one of bafflement' [*PLC,* 198].

Just as important in Richards's theory as this high estimate of poetry-reading's importance is the parallel elevation in status of 'the people who most value these experiences' — that is, critics. At one stage of his argument in *Principles of Literary Criticism,* he refers in passing to 'The people who are most keenly and variously interested, that is to say, the people whose lives are most valuable on our theory of value, the people for whom the poet writes and by his appeal to whom he is judged . . .' [*PLC,* 171]. It would appear that literary critics, and possibly a further small layer of discerning readers, are the most valuable people in society. This hint seems to be confirmed when Richards asserts: 'But it is not true that criticism is a luxury trade. The rear-guard of society cannot be extricated until the vanguard has gone further.' [*PLC,* 46.] It is in *Practical Criticism* that he seeks to address himself to the 'vanguard' of society — a small but possibly influential circle of English students and teachers in Cambridge. Although this work is best known for cataloguing the sources of bad reading and misguided criticism, Richards makes it clear that its major purpose is not to elaborate codes or principles of good criticism but to emphasize the crucial importance of the literary-critical act itself. This point is worth noting, against Richards's reputation as a highly theoretical system-maker. 'Thus no theory, no description, of poetry can be trusted which is not too intricate to be applied. . . . The choice of our whole personality may be the only instrument we possess delicate enough to effect the discrimination [between good and bad poetry].' [*PC,* 302.]

For Richards, systems and theoretical arguments about poetry and its value can only protect us against red herrings; they cannot themselves provide the basis for critical judgement. Such a basis exists only in the mobilization of the 'whole personality'. Critical theories are, indeed, a major obstacle to this kind of judgement:

And in general if we find ourselves, near this crucial point of choice, looking for help from arguments, we may suspect that we are on the wrong track . . . for it is in these moments of sheer decision that the mind becomes most plastic, and selects, at the incessant multiple shifting cross-roads, the direction of its future development.

The critical act is the starting-point, not the conclusion, of an argument. [*PC*, 303.]

Within the context of Richards's theory of value, which holds those experiences most valuable which enable the mind to incorporate most 'impulses' without confusion, this view of the critical act is of great importance. He is convinced that only a new flexible mind can face up to the problems of modern civilization and avoid chaos: a mind capable of loosening and recombining its constituent elements into new wholes, while retaining its balance. The arts, and poetry in particular, are held up as models of such a state of mind. But the real activity which is to keep alive this flexibility of mind is that of criticism or critical reading. Poetry itself is not the highest value, merely a storehouse or repository of value, an occasion for realizing value in the mind of a reader engaged in critical judgement upon it. It is the critical choice itself which imposes on us 'the supreme exercise of all our faculties' [*PC*, 305]. It entails a recasting of one's entire personality, a process which at first involves inner conflict but which has lasting rewards:

But when the conflict resolves itself, when the obstruction goes down or the crumple is straightened out, when an old habit which has been welcoming a bad poem is revivified into a fresh formation, or a new limb which has grown to meet a good poem wakes into life, the mind clears, and new energy wells up; after the pause a collectedness super-venes; behind our rejection or acceptance (even of a minor poem) we feel the sanction and authority of the self-completing spirit. [*PC*, 304.]

Criticism, by uniting all our faculties, seems to resolve tensions peaceably by bringing the mind into balance and even into 'self-completeness'. The significance of this last phrase, with its mystical overtones, is made apparent in the final paragraph of *Practical Criticism*, where Richards points out the need for a 'strengthening discipline' to equip us to withstand the pressures, internal and external, of the modern world.

The critical reading of poetry is an arduous discipline; few exercises reveal to us more clearly the limitations under which, from moment to moment, we suffer. But, equally, the immense extension of our capacities that follows a summoning of our resources is made plain. The lesson of all criticism is that we have nothing to rely upon in making our choices but ourselves. The lesson of good poetry seems to be that, when we have understood it, in the degree in which we can order ourselves, we need nothing more. [*PC*, 350–1.]

It is not entirely clear here whether he is saying that we need nothing more than ourselves or nothing more than good poetry. Either way the reader in communion with a good poem is regarded as 'self-complete', finished. Such a view, which pushes aside theories, preconceptions, and stock responses as machinery liable to obstruct the direct union of reader and poem, places a tremendous burden of responsibility upon the reader. If we have nothing to rely on in our choices but ourselves, then any choice we make about a poem will be at the same time an implicit judgement upon ourselves[20] — in fact a definitive (because 'self-complete') summary of our personality.

By stating that the basis for judging poetry is not theoretical considerations of value, but the whole personality of the critic, Richards helps to highlight the purpose of his own theory of value. Rather than aiding the judgement of one work of art against another, its function is to justify the elevation of poetry above other experiences and activities, stressing the mental incompleteness involved in every experience except that of great poetry. The relative value of particular poems is a question outside the scope of the general value-theory. Since the experience of poetry is 'self-complete', poetic value is its own proof: intelligent critics 'realise that no poem can be judged by standards external to itself' [*PC*, 243] — even standards arrived at via a value-theory. For Richards, 'value is prior to all explanations' [*PLC*, 135], and it 'cannot be demonstrated except by the communication of what is valuable' [*PC*, 12].

Although Richards may often appear to be proposing his theory of value as a framework within which various works can be placed according to merit, he actually comes out against any critical system 'external' to the poem. His real object of attention in the construction of the theory of value is the systematic establishment of relative merits not for poems but for minds or personalities, as he makes clear in an appendix to *Principles of Literary Criticism*: 'What the theory attempts to provide is a system of measurement by which we can compare not only different experiences belonging to the same personality but different personalities.' [*PLC*, 229.] The well-balanced mind to which poetry is the witness is taken as a touchstone against which different readers may be assessed in terms of relative mental health. How valuable one's personality is can be measured by the extent to which one pays attention to the self-validating virtues of great poetry. In Richards's view the primary problem facing criticism is not an abundance of bad poetry but the abundance of people who do not have the proper state of mind to recognize great poetry. The actual assessment of literary works and their likely effect is a matter of secondary importance:

The whole apparatus of critical rules and principles is a means to the attainment of finer, more precise, more discriminating communication. There is, it is true, a valuation side to criticism. When we have solved, completely, the communication problem, when we have got, perfectly, the experience, *the mental condition* relevant to the poem, we have still to judge it, still to decide upon its worth. But the later question nearly always settles itself; or rather, our own inmost nature and the nature of the world in which we live decide it for us. Our prime endeavour must be to get the relevant mental condition and then see what happens. [*PC*, 11.]

In accordance with this set of priorities, Richards sees his most urgent critical task as examining the various obstacles to the attainment by most readers of this appropriate mental attitude for poetry. The search is not for a poetry adequate to its age, but for what he calls an 'adequate reader' [*PC*, 109] fit to appreciate existing poetry. *Practical Criticism* is devoted to this task. The bulk of the book is taken up with an annotated anthology of comments ('protocols') made by students and other volunteers upon poems presented to them without the title or the author's name. Richards's reaction to examining these texts is consternation at his students' lack of self-reliance. 'Without further clues (authorship, period, school, the sanction of an anthology, or the hint of a context) the task of 'making up their minds about it', or even of working out a number of possible views from which to choose, was felt to be really beyond their powers.' [*PC*, 315.] The implied immaturity involved in such a failure is a constant target of reproach in *Practical Criticism*. Richards's comments on the wilful and Narcissistic disturbance of which bad poetry-reading is evidence have already been quoted above. They are echoed systematically in his remarks on the 'protocols' — as, for example, when a student criticizes a poem for its senti-mentality (misspelt as 'sentamentality'): '"Sentamentality" was certainly invited by the poem, and the invitation was not refused. As so often happens the reader's own revulsion at his own devious excesses is counted against the poet.' [*PC*, 88.] Once again Richards pits the abnormality of the reader against the supposed normality of the artist, on grounds no more substantial than amateur-psychoanalytic conjecture.

The consistency with which Richards attributes all 'failures in communication' in poetry-reading to various inadequacies and disturbances of the reader, rather than to the artist or to differences of history, language, and culture between the two, amounts to a systematic denigration of the reader. At one point Richards highlights this when he finds fault with the writers of the eighteenth century for paying their readers 'too much deference' [*PC*, 208]. He finds cause for regret in the lack of sanctions such as 'the penalty of being left out' for failures and errors in the use of language: 'A child of eight is constantly made to feel that he is not understanding something' [*PC*, 324], but older linguistic miscreants get away without even a sense of guilt. This denigration is not confined to the highly artificial conditions of the 'experiment', which denies to the reader all the bearings and connections within which poetry is actually read in the real world. (The peculiar conditions of the 'experiment' were exacerbated by a number of further omissions. From one poem Richards removed an entire line referring to the war, from another a stress mark whose omission rendered the sense quite ambiguous, and the title from an occasional poem. Nor could the temptations of context be entirely removed: the collocation of the poems in groups of four was assumed to be significant by many of the participants.)

The experiment, artificial as it is, is taken by Richards as hard evidence of his students' immaturity in life as a whole:

'How far can we expect such readers to show themselves intelligent, imaginative and discriminating in their intimate relations with other human beings?' [PC, 313]. One simple answer to this question is that they may be as intelligent, imaginative, and discriminating as Richards or anyone else, provided that they have the opportunity to know the names, backgrounds, and social circumstances of the human beings with whom they are intimate. Richards's unwarranted conclusion from the experiment has the basic flaw of failing to recognize that the unsigned and untitled poems he presented to the students were abstractions (that is, out of context) from the general life for which the students' responses allegedly made them so unfit. He seems to believe exactly the contrary: that the conditions of the experiment are more *real* in some sense than the world in which poets do publish more than one poem at a time (some even longer than fourteen lines) under their own names, in which literary schools and movements identify themselves, and reputations achieve currency. Richards's reasons for such a belief are important; they centre around ideas of self-reliance and personal independence. Under normal circumstances, he states, 'most of our responses are not real, are not our own' [PC, 349]. For him, the only 'real' response is an entirely individual one uncontaminated (as in a secret ballot, perhaps) by custom, reputation, fashion, debate, or — it seems — detailed background knowledge of what it is to which one is responding. The importance of an exercise which removes all these external props is that it puts the reader's entire personality to the test; the mind has nothing else to rely upon but its own resources:

Without *some* objective criteria, by which poetry can be tested, and the good distinguished from the bad, [the reader] feels like a friendless man deprived of weapons and left naked at the mercy of a treacherous beast. We decided that the treacherous beast was within him, that critical weapons — unless too elaborate to be employed — would only hurt him, that his own experience — not as represented in a formula but in its available entirety — was his only safeguard, and that if he could rely sufficiently upon this, he could only profit from his encounter with the poem. [PC, 314-5.]

With nothing objective against which to judge the poem, the reader has only himself or herself to bring to the poem as any kind of measure. But in Richards's theory there is no 'objective' poem either, only a mental response triggered by the poem. The result is that what is really tested is solely the personality of the reader, in a struggle with his or her own soul; ideally, in Richards's view, an inner striving for sincerity, self-recognition, and self-completeness, against the treacherous beast within.

The practical result of Richards's early work was the introduction of the 'practical criticism' exercise into the centre of the Cambridge English School's approach to the study of poetry — along with all its quasi-religious associations. Tillyard recalls that 'we felt rather as some of the first Protestants had felt about Scripture. This had been so overlaid with gloss and comment that the pure text had been hidden.'[21] The guarantee of this liberation of the text was the incorporation of Richards's method into the examination papers themselves. This, both in Richards's spiritually therapeutic sense and in the more mundane sense of academic assessment, was the real test: 'the greatest single achievement [of the school] . . . was to have introduced into the more advanced of the two purely literary sections a whole, compulsory, paper on practical criticism. . . . Here at last we could confront the men with the actual texts and test their ultimate literary insight, making them use their own resources entirely.'[22]

Whatever the truth of Richards's views on the efficacy of practical criticism for purposes of mental wholeness and sincerity, as a framework for the setting of examination questions it was a godsend. Moreover, as a technique for identifying the 'minds of the future', with their ability to accommodate several impulses or options without hastily acting on any of them, it was ideally suited to Cambridge's function in selecting and providing future senior civil servants. The old objection raised by E. A. Freeman to the establishment of an English school at Oxford ('As subjects for examination we must have subjects in which it is possible to examine') which had hung over the early English studies movement as a constant challenge to its stability as an academic discipline, could now be answered on its own terms. Avoiding mere Germanic 'fact-grubbing' on the one hand and vague impressionism on the other, the practical criticism examination offered the examiner a definite touch-stone (the pure text) against which candidates could readily be judged. It was, as Tillyard asserted, a foolproof test of the student's unaided perception: 'If properly set it cannot be faked . . .'.[23] The effect was widespread. Denys Thompson, in an essay entitled 'Teacher's Debt', attested to the radical change in Cambridge examination papers: 'A comparison of 1925 with 1963 shows that a knowledge of history, social background, linguistic origins and Aristotle had ceased to be tested.'[24] The Ministry of Education itself was to acclaim Richards's wider effect upon the teaching of English in schools throughout the country.[25]

iv. *'A rear-guard religious action'*

There have been many criticisms of Richards's theory and practice — most common among them the observation that the 'practical criticism' method is useful only with short poems, the charge that Richards excludes all historical and social factors in poetry, and that he overestimates poetry's effect upon the world. For the historian of literary criticism, at least, a more puzzling kind of attack requiring some investigation was that made upon Richards by T. S. Eliot in *The Use of Poetry and the Use of Criticism*.[26] For it had been Eliot's own work which had played an important role in bringing about the new practical criticism. As Tillyard recalls, Eliot's impact at Cambridge allowed Richards's project to take off as it did: 'the change of taste typified and promoted by Eliot, the reaction from Romantic emotionalism to more cerebral types of poetry, fostered the urge towards practical criticism because it directed attention to a kind of literature for which minute exegesis was especially apt. These circumstances combined to help Richards and his associates in what they had at heart.'[27] Richards, for his part, established himself in criticism partly by his willingness to go out and meet this disconcerting new literary figure. Undoubtedly his most far-sighted specific judgement upon a literary work was his recognition as early as 1926 of Eliot's *The Waste Land* as a major landmark in literature [*PLC*, 153 n., 231-5; *SAP*, 64-5]. But the considerations which Richards brought to bear upon that poem were a considerable embarrassment to its author. Not only did his recognition of the 'personal stamp' in every line of the poem challenge Eliot's notorious 'impersonality'; but his use of *The Waste Land* as an example of poetry which successfully cuts itself free from all beliefs troubled Eliot's new-found sense of the importance of religious orthodoxy.

Eliot met the remarks about his poem's freedom ·from all beliefs with an admission of perplexity. Richards's assertion that poetry can 'save us' was meaningless, Eliot complained, if we did not know what kind of salvation was intended. For Eliot, by this time an Anglo-Catholic, Richards's vagueness was worrying; it had no established system of beliefs behind it. He ridiculed what he saw as Richards's 'intense religious seriousness' about poetry.[28] The five subjects which Richards recommends that we contemplate deeply with our eyes closed, in order to enhance our sincerity of response, struck Eliot as shallow. On 'man's loneliness', Eliot remarks that he can understand the Christian sense of the separation of man from God, 'but not an isolation which is not a separation from anything in particular'.[29] He concludes that Richards's view of the collapse of beliefs or pseudo-statements is a pseudo-statement itself,[30] and Richards's work as a whole a poor substitute religion: 'I only assert again that what he is trying to do is essentially the same as what Arnold wanted to do: to preserve emotions without the beliefs with which their history has been involved. It would seem that Mr Richards, on his own showing, is engaged in a rear-guard religious action.'[31] From Eliot's point of view, such an artificial detachment of emotions was dangerous and confusing; it would disrupt the unity of sensibility normally guaranteed by religious orthodoxy. Poetry could not save us, either spiritually or as an antidote to future social catastrophes. For both purposes, a rock of religion was required at the centre of society's culture.

Richards's views were damaging not to religion alone, in Eliot's opinion, but to poetry too. Eliot observes, in connection with the problem of 'difficult' poetry, that readers who are led to expect and to seek complexity in poetry will suffer in consequence:

The ordinary reader, when warned against the obscurity of a poem, is apt to be thrown into a state of consternation very unfavourable to poetic receptivity. Instead of beginning, as he should, in a state of sensitivity, he obfuscates his senses by the desire to be clever and to look very hard for something, he doesn't know what — or else by the desire not to be taken in. There is such a thing as stage fright, but what such readers have is pit or gallery fright.[32]

Although Richards is not named here, it is unlikely that his emphasis upon poetic complexity was far from Eliot's mind when this was written. The argument, from a writer much of whose earlier criticism was devoted to justifying difficult, allusive poetry, seems to represent a stepping-back from conclusions which could be drawn from his own work. As Eliot found them reflected in Richards's writings, the themes of *The Sacred Wood* and of the essays of the early 1920s had taken on a transformed life of their own, increasingly out of step with his own evolution. For Richards had answered Eliot's appeal for a criticism which took 'the side of the artist' — to such an extent that the reader was now virtually intimidated. He had taken Eliot's distinction between the poet and the thinker to the extent of denying the need for any beliefs at all in poetry. He had adopted Eliot's ideal of the unified sensibility, turning it into an absolute criterion of all value. When Eliot recoiled from many of the critical ideas propounded by Richards, he was reacting to what in many ways was a brainchild of his own, now being led on an increasingly divergent path. The divergence was, at bottom, a difference of approach to the common aim of preserving the fabric of society and its traditional cultural standards. Where Arnold and Richards advocated a policy of ideological flexibility whereby a minority culture could absorb or ride over the mounting threats to its cohesion, Eliot began to insist that only a policy of retrenchment could preserve the integrity of traditional values. Richards captured the distinction vividly when he offered the alternative of a rock to cling to in the coming storms or an aeroplane to ride them out. Eliot's final choice was the institutional rock of the Church, to which he paid tribute in his religious verse.

Richards drifted. After completing *Practical Criticism* he seems gradually to have lost interest in Cambridge, and set out to pursue his hopes of world peace through improved linguistic communication — elaborating the hopes expressed in the Newbolt Report for English as a world language. What he left behind him for others to develop into 'Cambridge English' was a conception of the study of literature as an indispensable agent of social cohesion, combined with a teaching and examining method — 'practical criticism' — which could at last establish this conception in institutional form.

NOTES

1. E. M. W. Tillyard later admitted: 'Neither Forbes, nor Richards, nor I would have stood the least chance of getting on the English staff if there had been an English Appointments Committee operant at the dates when Chadwick told the Special Board to put us on the lecture list.' *Muse Unchained*, 111.

2. Reuben Brower, Helen Vendler, John Hollander (eds.), *I. A. Richards: Essays in His Honour* (1973), 233.

3. I. A. Richards, *Complementarities*, ed. J. P. Russo (Manchester, 1977), 257.

4. Tillyard recalls: 'In the third year of the war thoughtful people began to think ahead about conditions after it, which they knew would be different from any that had gone before. They knew that war was in itself an agent of revolution and that the men who had served abroad would not be satisfied with the narrowly linguistic cast of the old examinations.' Tillyard, *Muse Unchained*, 58.

5. Brower *et al.*, *Richards*, 257.

6. Ibid., 73.

7. C. K. Ogden and I. A. Richards, *The Meaning of Meaning* (1923), 17.

8. Ibid., 18n.

9. *SAP*, 35. In Richards's later writings, particularly *Speculative Instruments* (1955), the metaphor of the United Nations is constantly employed for the peaceful arrangement of the 'impulses' and for philosophy's reconciliation of academic disciplines.

10. Richards recommended the Report highly, adding that Sampson's *English for the English* 'should on no account be overlooked' [*PC*, 333n].

11. *PLC*, 25-6. Richards remarks later in the same work that the cinema is a medium 'that lends itself to crude rather than to sensitive handling' [*PLC*, 182]. When attributing immature attitudes to 'bad literature, bad art, the cinema, etc.' [*PLC*, 159], he appears not to have entertained the possibility of good cinema.

12. *PLC*, 26. Cf. Quiller-Couch on the 'aristocratic' view of taste: 'Such an attitude towards any form of art seems to me, Gentlemen, morose and out of place in our Commonwealth. It inclines the patient, for example, to say of any exotic statue offensive to the public, "stand aside, or pass it with acceptance, until you have learnt that it is good for you. *We* are the *cognoscenti* and proclaim it to be good without giving reasons which you would not understand." Now really this will not do for a well-ordered State.' *Studies in Literature: Third Series* (Cambridge 1929), 194 ('On "The New Reading Public", I').

13. A similar view entertained by Arnold seems to foreshadow the 'practical criticism' exercise: 'It would be extremely interesting to make a raid among the youth of the wealthier classes, whether at their schools and universities, or at their scenes of amusement, to catch five or six hundred of them from the age of eighteen to that of twenty-five and to subject them to the same test of their general intelligence to which, by this passage of poetry to be paraphrased, the general intelligence of the candidates from elementary schools is subjected.' *Reports on Elementary Schools 1852-1882*, ed. Sandford, 178-9.

14. *PLC*, 191. The precise meanings of the terms 'impulse' and 'experience' are never clear in Richards's writings. They seem to undergo significant shifts according to the occasion, enabling Richards to slide easily over some of the thorniest problems in psychology. See W. H. N. Hotopf, *Language, Thought and Comprehension* (1965), 310-12, and John Needham, *The Completest Mode* (Edinburgh 1982), 31-5.

15. Richards's later claim that he 'didn't . . . know anything about Coleridge' [*Complementarities*, 257] until after the writing of *Practical Criticism* is not to be taken seriously. Some of his earliest remarks on aesthetic 'equilibrium' — that it 'tends to bring the whole personality into play', or that it 'brings into play all our faculties' [C. K. Ogden, I. A. Richards, and James Wood, *The Foundations of Aesthetics* (1922), 77, 91], are taken directly from Coleridge, to whom he pays explicit tribute elsewhere [*PLC*, 108, 191].

16. *SAP*, 58-9. Cf. Eliot: 'A philosophical theory which has entered into poetry is established, for its truth or falsity in one sense ceases to matter, and its truth in another sense is proved.' *SE*, 288-9.

17. *The Foundations of Aesthetics*, 78.

18. *Speculative Instruments*, 44n.

19. *PLC*, 22-3. Cf. the Newbolt Report's definition of literature as 'the record of the experiences of the greatest minds', *The Teaching of English in England* 149. Behind both formulations lies Shelley's 'A Defence of Poetry', where poetry is described as 'the record of the best and happiest moments of the happiest and best lives', and poets as 'spirits of the most refined organisation', *Prose Works*, ed. R. H. Shepherd (1912), ii. 33-4.

20. Sainte-Beuve, preoccupied with the nature of 'judgement', had formulated the same idea: 'Et en général nos jugements nous jugent nous-mêmes bien plus qu'ils jugent les choses.' (And in general, our judgements pass judgement upon us ourselves far more than they judge

other things.') *Correspondance générale*, vi, ed. Jean Bonnerot (Paris, 1949), 272 (6 November 1845).

21. Tillyard, *Muse Unchained*, 82-3.

22. Ibid., 108-9.

23. Ibid., 137.

24. Brower *et al.*, *Richards*, 259.

25. An official pamphlet stated that Richards's work 'more than any other single influence ... has helped to change the spirit and method of study of poetry in grammar schools and therefore indirectly in all schools', *Language. Some Suggestions for Teachers of English and Others* (Ministry of Education Pamphlet no. 26, HMSO 1954), 145.

26. See also Eliot's remarks in 'A Note on Poetry and Belief', *The Enemy*, i. (1927), 15-16, and 'Note' to 'Dante', *SE*, 269-71.

27. Tillyard, *Muse Unchained*, 100.

28. Eliot, *The Use of Poetry and the Use of Criticism*, 132.

29. Ibid.

30. An opinion endorsed by William Empson in *The Structure of Complex Words* (1951), 422-4.

31. Eliot, *The Use of Poetry and the Use of Criticism*, 134-5.

32. Ibid., 150-1.

7. THE LEAVISES: ARMED AGAINST THE HERD

The new-comer already threatened to be a nuisance in
the shape of rivalry, and was certainly a nuisance in
the shape of practical criticism . . .
George Eliot, *Middlemarch*

When T. S. Eliot wrote, in *The Use of Poetry and the Use of Criticism,* that 'every hundred years or so, it is desirable that some critic shall appear to review the past of our literature, and set the poets and the poems in a new order',[1] that reordering, or 'revaluation' as it became known, was already under way. While the early promise of both Eliot and Richards was rapidly drying up, two younger critics at Cambridge — F. R. and Q. D. Leavis — were laying the foundations for a literary criticism which would dominate its century as Dryden, Johnson, and Arnold had theirs; a project which was not only to 'revalue' the accepted order of English poetry, but to establish the first authoritative critical scale upon which the English novelists were to be measured. Their critical career, spanning over forty years and including the production under their guidance of *Scrutiny* for twenty of those, was inaugurated with the publication between 1930 and 1933 of a series of cultural manifestos: *Mass Civilisation and Minority Culture* (1930), *Fiction and the Reading Public* (1932), and *Culture and Environment* (1933). These works, together with some early essays by F. R. Leavis collected along with *Mass Civilisation and Minority Culture* in *For Continuity* (1933), formed the founding documents of a remarkably coherent literary-critical movement; documents rich enough in their literary and sociological implications to sustain a growing corps of teachers and students for decades to come in further elaboration. It is these texts which map out the social, historical, and ethical assumptions without which the necessary consensus implicit in, say, *The Great Tradition* (1948) or much of *Scrutiny* — the 'we' who partake in their literary judgements — would not be possible. Even a 'pure'

literary-critical work of the same period, such as F. R. Leavis's
New Bearings in English Poetry (1932), depends on these
texts for its own 'bearings': its argument is framed at each
end, by an introductory restatement of Richards's view of the
poet as 'unusually sensitive, unusually aware, more sincere
and more aware than the ordinary man can be',[2] and by an
epilogue again resuming Richards's analysis of the mass-
produced, standardized, and levelled-down context of modern
literature. It is the manifestos, where these ideas are concentrated most clearly, that this chapter will examine.

i. *An anti-Marxist consensus*

In *Mass Civilisation and Minority Culture* F. R. Leavis at
once takes up the defence of threatened minority values
where Arnold and Richards had left off, remarking upon the
deteriorating cultural conditions for such a defence compared
with the task Arnold had faced in his day. Not only have
the machine, the cinema, and their attendant effects of
standardization advanced, but the very terms of the resistance,
the language of Culture (Leavis cites 'the will of God' and
'our true selves'), can no longer be taken for granted as
Arnold had taken them. That certain assumed traditional
values had always been confined to a minority was inevitable
and held no great danger in itself, so long as these values
commanded some kind of deference from the majority.
But what was new and threatening in the post-war world
was precisely that the 'mass' was beginning actively to
challenge the status of the minority, creating an oppositional
language subversive of cultural authority. The appearance of
the word 'high-brow' is identified by Leavis as the most
alarming evidence of this trend: '"High-brow" is an ominous
addition to the English language. I have said earlier that
culture has always been in minority keeping. But the minority
now is made conscious, not merely of an uncongenial, but of
a hostile environment.' [*FC*, 38.]

While Shakespeare 'appealed at a number of levels of
response, from the highest downwards' [*FC*, 38], and Fielding
and even Hardy had been able to command such a mixed
audience, today there was a divorce between the 'levels' of

Eliot, Woolf, Joyce, or Pound, and the work of Wells or Bennett, or, lower still, Edgar Rice Burroughs's *Tarzan of the Apes*. The attitude implicit in the use of the word 'highbrow' is, for Leavis, as much a cause of this divorce as an effect. If modernist literature is accessible only to a minority, it is driven into this isolation by a hostile environment. 'The minority is being cut off as never before from the powers that rule the world. . . . 'Civilisation' and 'culture' are coming to be antithetical terms. It is not merely that the power and the sense of authority are now divorced from culture, but that some of the most disinterested solicitude for civilisation is apt to be, consciously or unconsciously, inimical to culture.' [*FC*, 39.] Particularly worrying in this respect was the desertion, as it must have seemed, of no less a figure than I. A. Richards to the cause of Basic English — a project which ultimately threatened the subtlety of the language upon whose fate depended (as Richards himself had pointed out) the entire cultural heritage. If Richards could not foresee the consequences of his misguided concern for literacy, then it fell to his followers to remind him and the minority he had defended of the urgent case presented in *Science and Poetry*.

Comparing the very small minority capable of first-hand judgement of art and literature with a small proportion of gold upon which a paper currency is based, Leavis quotes Richards's defence of criticism against the charge of being a luxury trade — the passage in which he conceives of criticism as a vanguard of society, a guardian of mental health, and in which the artist is characterized as 'the man who is most likely to have experiences of value to record' [*PLC*, 46]. Leavis extends Richards's claims for criticism thus:

The minority capable not only of appreciating Dante, Shakespeare, Donne, Baudelaire, Hardy (to take major instances) but of recognising their latest successors constitute the consciousness of the race (or of a branch of it) at a given time. For such capacity does not belong merely to an isolated aesthetic realm: it implies responsiveness to theory as well as to art, to science and philosophy in so far as these may affect the sense of the human situation and of the nature of life. Upon this minority depends our power of profiting by the finest human experience of the past; they keep alive the subtlest and most perishable parts of tradition. Upon them depend the implicit standards that order the finer

living of an age, the sense that this is worth more than that, this rather
than that is the direction in which to go, that the centre is here rather
than there. In their keeping, to use a metaphor that is metonymy also
and will bear a good deal of pondering, is the language, the changing
idiom, upon which fine living depends, and without which distinction
of spirit is thwarted and incoherent. By 'culture' I mean the use of such
a language. I do not suppose myself to have produced a tight definition,
but the account, I think, will be recognised as adequate by anyone who
is likely to read this pamphlet. [*FC*, 14-15.]

Three points in this statement deserve particular notice.
First, the identification of the literary minority as the consciousness of the race is not qualified to any degree by the
apparent concessions made to science and philosophy; for
their claims demand a response from the minority only 'in so
far as [they] may affect the sense of the human situation. . .'.
As Arnold had done in 'Literature and Science', Leavis implies
a certain inhumanity in science and philosophy, leaving them
outside the human sense accessible to the literary critic. It is
those who appreciate Dante, Shakespeare, Donne, and
Baudelaire who set the terms upon which, say, Aristotle,
Darwin, Marx, Freud, or Einstein and their latest successors
may or may not be admitted to the consciousness of the race.
What is conceded here to science with one hand is taken back
with the other. The second noteworthy point is the attribution
to the minority of readers and critics of qualities originally
identified by Richards (and by Ezra Pound) with the peculiar
qualities of the poet. Although it was noted in the previous
chapter that Richards included critics as well as poets in the
vanguard of the species, Leavis develops this hint well beyond
the boundaries of Richards's usual aesthetic and psychological
concerns. Leavis is preoccupied more with this enlarged
vanguard's cohesion, its effectiveness as a social force, and its
degree of influence upon those who have power. That he is
addressing an audience within the minority should be evident
from the third notable feature of the passage: its appeal to
those 'likely to read this pamphlet'. For this readership, tight
definition of the values of 'culture' is unnecessary, or if any
of them feel that it is necessary, they cannot be counted
among those who share 'implicit standards' with Leavis. *Mass
Civilisation and Minority Culture,* which was first published

by the Minority Press, is not intended to impress others with the importance of the literary minority, but to re-establish among the minority itself its duty not to disintegrate under the mounting pressures of civilization.

Identifying and rallying the vanguard required at the same time surveying the destructive forces arrayed against it. The general symptoms are those familiar to Richards's readers: standardization, stock responses, levelling down. Where Leavis develops the diagnosis is in placing less emphasis on the growth of population and more on 'the machine': the automobile, for example, has broken up the family so catastrophically that 'parents are helpless to deal with their children' [FC, 17]. An additional element in Leavis's diagnosis is his approach to the stock response as a deliberate exploitation rather than as natural inertia. Richards had warned that 'we have not yet fathomed the more sinister potentialities of the cinema and the loudspeaker'. One of Leavis's aims is to correct this gap in our knowledge. The cinema, he asserts, involves 'surrender, under conditions of hypnotic receptivity, to the cheapest emotional appeals' [FC, 21], and the wireless finds its audience in the same enfeebled state: 'But perhaps it will not be disputed that broadcasting, like the films, is in practice mainly a means of passive diversion, and that it tends to make active recreation, especially active use of the mind, more difficult. And such agencies are only a beginning. The near future holds rapid developments in store.' [FC, 21-2.]

The most ominous of these developments is the increasingly sophisticated use of market research methods by advertising agencies to perfect the stock responses of the masses. The necessary complement to the helpless passivity or suggestibility of the mass audience is the growing astuteness and expertise of the professional suggestors. 'Contemplating that deliberate exploitation of the cheap response which characterises our civilisation we may say that a new factor in history is an unprecedented use of applied psychology.' [FC, 22.] Leavis cites representatives of the advertising profession confidently predicting the advent of perfected consumer demand manipulation. Worse still, the skills of the copywriter are now extending their influence into the world of literature itself, as writers like Arnold Bennett prostitute their critical discrimination to the marketing requirements of the Book Society Ltd. This analysis is only a preliminary to the Leavises' more extended critique of advertising which will be discussed in more detail below. What immediately concerns Leavis at this point is the threat posed by advertising, and by Basic English, to the language upon which fine living depends: a theme developed, as many of the suggestions in *Mass Civilisation and Minority Culture* were to be, in the series of early essays collected along with it in *For Continuity*.

The claims which Leavis makes for the high status of the critical minority as keepers of the language and preservers of the tradition are elaborated substantially in 'The Literary Mind' by appealing to the authority of a crucial passage in Ezra Pound's *How to Read* (1931):

'Has literature a function in the state, in the aggregation of humans, in the republic, in the *res publica* . . .? It has. . . . It has to do with maintaining the very cleanliness of the tools, the health of the very matter of thought itself. Save in the rare and limited instances of invention in the plastic arts, or in mathematics, the individual cannot think and communicate his thought, the governor and legislator cannot act effectively or frame his laws, without words, and the solidity and validity of these words is in the care of the damned and despised *literati*. When their work goes rotten — by that I do not mean when they express indecorous thoughts — but when their very medium, the very essence of their work, the application of word to thing goes rotten, i.e., becomes slushy and inexact, or excessive or bloated, the whole machinery of social and of individual thought and order goes to pot.' This is well said. Literary criticism has a correspondingly high function, and literary study . . . should be the best possible training for intelligence — for free, unspecialised, general intelligence, which there has never at any time been enough of, and which we are peculiarly in need of to-day. [*FC*, 53-4.]

Pound's argument, partly by virtue of its vigorous hygienic metaphor, amounts to a much less 'slushy and inexact' restatement (but a restatement none the less) of Shelley's case for regarding the poet as an unacknowledged legislator. It can easily be seen how a wholehearted acceptance of the argument leads directly to a vertiginous sense of responsibility in the critic, who now becomes a fine thread upon which the fate of governments and civilizations dangle. Furthermore,

Pound's explicit connection between the health of literature
and the health of the state leads Leavis in a direction only
glancingly hinted at in *Mass Civilisation and Minority Culture,*
where he had spoken of the minority's isolation from the
centres of power. In another essay, entitled 'What's Wrong
with Criticism?', he draws attention to the paradox that
contemporary criticism has reached a remarkable high point
of achievement, in the work of Eliot and Richards, while at
the same time sinking into virtual insignificance as a social
force: 'a consciousness maintained by an insulated minority
and without effect upon the powers that rule the world has
lost its function' [*FC*, 72]. No amount of technical expertise
of the kind represented by Richards can restore criticism to
a position of influence, for the problem is not one of individual
sensibility (important as that is in Leavis's view) but of a
whole culture, and to tackle a whole culture a sense of tradition
transcending the individual is necessary. What Leavis is proposing, when he asserts that 'Sensibility and the idea of
tradition — both concerns are essential' [*FC*, 66], is a combination of the major concerns of Eliot and of Richards, but
now with a specific aim: to establish criticism as a social force.

If criticism is to become a power in the land, it must obtain
points of support in society — centres of authority around
which a scattered literary minority can regroup and from
which it can reassert its standards. Surveying the possible
cultural authorities which could assume the role of an
Arnoldian academy, Leavis dismisses the Royal Society of
Literature, the English Association, and the BBC as bodies
irreparably corrupted by the ethos, if not the direct interference, of the Book Society Ltd. He concludes 'What's
Wrong with Criticism?' (an essay commissioned by Eliot for
the *Criterion* but then rejected) on a note of resolute pessimism,
comforting himself only in the fact that 'there are some to
whom the substance of this essay is commonplace, otherwise
it would not have been worth writing' [*FC*, 90]. Leavis claims
to have convinced no one; the immediate problem is to
evince mutual recognition from those who already know, and
to begin organizing. But it was not as if, with the failure of
the Royal Society, the English Association, and the BBC, all
avenues to influence had been foreclosed. A more fruitful

direction was indicated in 'The Literary Mind', where Leavis concedes that 'luck' might provide 'a centre of stimulus and a focus of energy at some university' [FC, 65], and then goes on to suggest possibilities for the teaching of English in schools to be tied to the study of the sociological and cultural context of literature. If a real training in sensibility were to be begun in schools, 'a good deal might be done to cultivate a critical awareness of contemporary civilization' [FC, 66].

This tentative placing of hopes in the educational system becomes, in the later essay 'Restatement for Critics', a much firmer commitment, a choice made or precipitated under the pressure of Marxist objections to the ambiguities of Scrutiny's aims. Against charges of complicity in the maintenance of bourgeois culture, Leavis reaffirms his choice of education as the appropriate terrain of resistance to the established system, considerably stressing the radical content of his educational proposals:

Whether or not we are 'playing the capitalist game' should soon be apparent, for a serious effort in education involves the fostering of a critical attitude towards civilisation as it is. Perhaps there will be no great public outcry when it is proposed to introduce into schools a training in resistance to publicity and in criticism of newspapers — for this is the least opposable way of presenting the start in a real modern education. Yet the inevitable implications, accompaniments and consequences of such a training hardly need illustrating.

The teaching profession is peculiarly in a position to do revolutionary things: corporate spirit there can be unquestionably disinterested, and by a bold challenge there, perhaps the self-devotion of the intelligent may be more effectively enlisted than by an appeal to the Class War. [FC, 188-9.]

Although Leavis had made it clear that the crucial problem of culture was the separation of the literary minority from power and public authority, as a means of resolving the problem he steers a deliberate course away from political partisanship and towards stealthy educational reform.

Several of the essays collected in For Continuity bear witness to the importance of Marxism as a 'whetstone' against which Leavis was obliged to define and focus his positions. As a movement competing for the loyalties of many of the disaffected young intellectuals whom Leavis himself was trying to address, it challenged him to elaborate his analysis of civilization and to specify his remedies. Though Leavis refuses, in '"Under Which King, Bezonian?"', to make any positive commitment to one political cause or another, his answer to the claims of Marxism — in the negative — emerges clearly enough from all these essays. Perhaps in an attempt to outbid the radicalism of the Marxists, he not only envisages 'revolutionary' possibilities for the teaching profession, but accuses his left-wing critics of being just as bourgeois as their apparent enemies:

Class of the kind that can justify talk about 'class culture' has long been extinct. . . . The process of civilisation that produced, among other things, the Marxian dogma, and makes it plausible, has made the cultural difference between the 'classes' inessential. The essential differences are indeed now definable in economic terms, and to aim at solving the problems of civilisation in terms of the 'class war' is to aim, whether wittingly or not, at completing the work of capitalism and its products, the cheap car, the wireless and the cinema. [*FC*, 172.]

While recognizing that economic class divisions exist, Leavis insists that capitalist civilization is *culturally* homogeneous: the cinema-going masses and the advertising magnates reinforce their shared values in unbroken consensus, just as the Marxist and the bourgeois both connive at the destruction of tradition by the machine. As part of this established consensus, Marxists are unqualified to speak of cultural values; these can only be renewed by a consensus of a different kind, drawn from outside the common assumptions of the bourgeois-Marxist bloc. However, Leavis does say, in what has often been mistaken for a concession, that Marxists are qualified to speak with authority on economic, rather than cultural questions. As he implies in the passage above and repeats in the preface to *For Continuity,* 'the only real differences between classes today are economic'.[3] In this limited realm, the Marxists may be allowed as much of the truth as they care to claim, since for Leavis the real problem lies elsewhere:

Let me say, then, that I agree with the Marxist to the extent of believing some form of economic communism to be inevitable and desirable, in the sense that it is to this that a power-economy of its very nature points, and only by a deliberate and intelligent working towards it can civilisation be saved from disaster. (The question is, communism of what kind? Is the machine — or Power — to triumph or to be triumphed over, to be the dictator or the servant of human ends?)[4]

Where Leavis appears most to agree with Marxism, as here, he formulates most exactly, by an apparent paradox, his fundamental divergence from it. For the repeated specification of Marxism's proper realm (he refers again in '"Under Which King, Bezonian?"' to 'communism as the solution to the economic problem') effectively quarantines it in a way which leaves the problem of 'Power' in its other sense quite untouched. Leavis's ready agreement with a purely economist caricature of 'Marxism' smothers the essential Marxist assertion that class differences extend beyond the economic and into the political realm, i.e. that one class wields the power of the state 'machine' while the others do not. With the concept of power confined in this way to an economic and technological reference, Leavis's concession amounts to a kiss of death for his Marxist opponents.[5] If the focus of power in society is the machine and not the state, there can be no essential conflict between the Marxist and the bourgeois, committed as they are to the extension of an identical power. Politics, as such therefore ceases to exist for Leavis; there remain only the realm of economics and the realm of culture.

The Marxists, conceived of in this way as part of an 'economic' alliance, are accordingly not the sole targets of Leavis's campaign against the 'economic' view of life. Several times he attempts to reclaim the sense of the 'standard of living' as the quality of life rather than as a quantitative measure of machine-products; he speaks of his effort to 'wrest the phrase from the economist' [FC, 17]. For Leavis, the values of humane letters touch upon the ends of human life, while science and economics consider only the means, the machinery. This kind of narrow concern for the 'standard of living' only in its economic sense condemns not just the Marxists but such prominent figures in British culture as Bernard Shaw, Bertrand Russell, and H. G. Wells [FC, 63]. A repeated accusation, made against Wells, Marx, and the left-wing American novelist John Dos Passos, is that they assume art and literature and their associated values to be capable of looking after themselves once the economic problem is solved: 'To hope that, if the mechanics of civilisation (so to speak) are perfected, the other . . . problems will solve themselves, is vain.'[6] But Leavis can be accused, with equal justice, of hoping that if the 'non-economic' and 'non-political' problems of life were approached directly and sincerely, everything else would look after itself. He quotes approvingly from D. H. Lawrence: '"They simply are," he wrote, "so eaten up with caring. They are so busy caring about Fascism or Leagues of Nations or whether France is right or whether Marriage is threatened, that they never know where they are. They certainly never live on the spot where they are." Lawrence always lived on the spot where he was. That was his genius.' [FC, 153.]

Lawrence's genius, which Leavis sees as a unique sign of hope in an otherwise stiflingly mechanized culture, is a question not just of being 'on the spot', wherever that might be, but of being rooted to it. By this it is meant, first, that Lawrence's 'gift lay, not in thinking, but in experiencing', that he had a genius for 'particular experience' [FC, 57, 59] from which abstract doctrines are but a distraction; and, in addition, that this gift puts him in touch with a tradition inaccessible to those obsessed with economics, machinery, and abstractions: 'it was in the past that he was rooted. Indeed, in our time, when the gap in continuity is almost complete, he may be said to represent, concretely in his living person, the essential human tradition; to represent in an age that has lost the sense of it, human normality, as only great genius could.' [FC, 158.] Richards's doctrine of the normality of the artist is reinforced when Leavis writes of Lawrence as 'normal, central and sane to the point of genius, exquisitely but surely poised . . .' [FC, 152]. For Leavis, these virtues, combining Ricardian openness to experience with Eliotic sense of tradition, are the virtues required in any movement to resist the inhuman tendencies of the modern age.

As Leavis puts it in 'The Literary Mind', no social or political movement could engage his faith and energy unless it were related to a desperate attempt to revive or replace a decayed tradition. If all that the bourgeois or the Marxist could offer was a future of increasingly perfected machinery,

then that future was not worth the having; something more
appealing was to be recovered from the past. In John Dos
Passos's *U.S.A.* trilogy Leavis recognizes an impressive
depiction of the moral decay of contemporary civilization,
but notes that Dos Passos has failed to locate any focus of
renewal other than hints of social revolution. For Leavis, this
represents a failure to grasp the extent of the decline: 'What
has disintegrated — this is the point — is not merely "bourgeois" or "capitalist" civilisation; it is the organic community.'
[*FC*, 108.] No future can be attractive which does not
embody the values of the pre-mechanical age. 'The memory
of the old order, the old ways of life, must be the chief hint
for, the directing incitement towards, a new, if ever there is
to be a new. It is the memory of a human normality or
naturalness . . .' [*FC*, 109]. It is here that the 'continuity'
aspired to in the title of the collection takes on its significance.
Although Leavis does not specify the nature of the organic
community, beyond alluding to Bunyan and George Bourne,
he makes clear that any movement capable of influencing
the future must be, like Lawrence, rooted in the past. And,
like Lawrence, it must find these roots by cultivating a sense
of experience, of the concrete; a sense which literary criticism
is uniquely able to instil. 'To be concerned, as *Scrutiny* is,
for literary criticism is to be vigilant and scrupulous about
the relation between words and the concrete', while the
Marxists are guilty of 'shamelessly uncritical use of vague
abstractions and verbal counters' [*FC*, 171].

Scrutiny, then, is to take upon itself the preservation of
linguistic health, the cleanliness of the very tools of thought,
in Pound's sense. To carry out this task, it must avoid any
taking of sides in the phoney war of abstractions waged
between different political movements. To identify *Scrutiny*
with any creed would, as Leavis writes in '"Under Which
King, Bezonian?"', only compromise its special function,
distracting it and its readership from the immediate relations
of language and concrete experience. For a renewal of
traditional values must start, as it were, from the ground
upwards — a surer method than the building of ideological
castles in the air. The clearest explanation of this method
appears in 'Restatement for Critics':

And we know that, in such a time of disintegration as the present, formulae, credos, abstractions are extremely evasive of unambiguous and effective meaning, and that, whatever else may be also necessary, no effort at integration can achieve anything real without a centre of real consensus — such a centre as is presupposed in the possibility of literary criticism and is tested in particular judgements . . .

The peculiar importance of literary criticism has by now been suggested; where there is a steady and responsible practice of criticism a 'centre of real consensus' will, even under present conditions, soon make itself felt. Out of agreement and disagreement with particular judgements of value a sense of relative value in the concrete will define itself and, without this, no amount of talk about 'values' in the abstract is worth anything. [*FC*, 182–3.]

The 'restatement' here is a résumé not just of Leavis's positions, but of Arnold's views on literary education: the judgements which form themselves without any turmoil of controversial reasoning, the belief that poetry is the reality and philosophy the illusion (Leavis, indeed, had recalled this formula of Arnold's when discussing Wordsworth and Lawrence in 'The Literary Mind'). Leavis goes further, however, in developing the concept of the consensus which is implied in the very practice of literary criticism. It was to become a repeated message of Leavis's writings that literary criticism makes its judgements in a corroborative mode — 'This is so, is it not?' The task of *Scrutiny* was to extend this small-scale intimacy of reader and writer, or of student and teacher, into a larger consensus embracing an entire literary minority, reorganizing it upon the basis of tested values. The consensus is already there, as Leavis insists so often in these essays; his dialogue of the deaf with the Marxists is really intended only for the ears of those for whom his arguments are already obvious. The aim is for Arnold's 'remnant' not so much to convince anyone else as to recognize one another and to organize themselves into an effective force.

If a new movement to combat the disintegration of civilization had to be 'rooted' as Lawrence was to the solid ground of concrete experience, this did not at all imply that the recommended 'living on the spot' was to be a form of 'living in (or for) the present'. On the contrary, the present was too contaminated for any roots to be put down here. Shakespeare had had the advantage of a 'national culture rooted in the soil' [*FC*, 216] which was now simply unavailable to the contemporary urban writer except via the literary traditions of the past. While the contemporary culture available to Shakespeare was a source of strength, that available to the modern writer can only be an impediment. The point is made most concisely in Leavis's discussion of *Literature and Revolution*, the work of 'that dangerously intelligent Marxist' Trotsky. When Trotsky insists upon the preservation of the essential elements of bourgeois culture, against proposals for the creation from scratch of a completely new 'proletarian culture', Leavis approves, adding: 'The problem is, rather, not merely to save these 'essential elements' from a swift and final destruction in the process that makes Communism possible, but to develop them into an autonomous culture, a culture independent of any economic, technical or social system as none has been before.' [*FC*, 167–8] Leavis describes this as 'a rootless culture' — an apparently paradoxical reversal of the positive value he usually attributes to 'rootedness'. The paradox is explicable only in terms of 'tradition' as Eliot had presented it: a continuity outside that of real historical time. In this view of tradition, the real present can be absolutely discontinuous with the past. To preserve continuity, then, the new autonomous culture which Leavis projects has to be rooted in the past, but just as emphatically uprooted from the present. *For Continuity* continually reiterates the need for such a disengagement, but only glancingly alludes to the roots in the past which Leavis wishes to establish. This encounter with history is left for fuller treatment in *Fiction and the Reading Public* and *Culture and Environment*.

ii. *Sociology of the herd*

I. A. Richards had been Q. D. Leavis's supervisor at Cambridge, and it is his work which is acknowledged as her inspiration in the introduction to *Fiction and the Reading Public* — in particular the remark he had made in *Principles of Literary Criticism* that criticism needed to give clear reasons why those who scorned bestsellers were not necessarily snobs. *Fiction and the Reading Public* is partly an attempt to extend

Richards's methods into the neglected field of prose fiction, yet in addressing the wide questions of snobbery and of critical values it has to be more than this. Leavis finds the usual method of academic literary criticism unsuited to her purpose, since she is dealing with 'fiction as distinct from literature' [*FRP*, 13], while the purely scholarly approach of cataloguing plots and characters evades the question of value. The result is the adoption of a third approach, which she calls the 'anthropological' (or at times 'sociological') method: examining all the relevant material and allowing conclusions 'to emerge simply by comparison and contrast and analysis' [*FRP*, 14]. This effort to present judgements as making themselves without the interference of authorial dogma is familiar, but the disclaimer is rapidly and amply refuted by *Fiction and the Reading Public* itself. It is clear that Leavis does have a coherent case to argue, and that her material is presented according to a definite historical and sociological model. Indeed it is the great merit of the book that, for the first time since *Culture and Anarchy*, it attempts extensively to explain from the critic's point of view the world in which literary culture is to intervene. In a critical tradition whose view of the world is so often only 'implicit', such a work is a rare opportunity to survey the bases of the values deployed.

Where T. S. Eliot had diagnosed a 'dissociation of sensibility' in English culture (or the 'English mind') since the seventeenth century, Q. D. Leavis seeks a more tangible development, transferring Eliot's view of history to some extent out of the realm of mind or sensibility and into identifiable social groups and institutions. In her more detailed account of cultural decline it is above all a unitary reading public which disintegrates, and with it a coherent 'society' or community of taste. This is an important advance upon Eliot's original thesis — an advance in the direction of historical credibility and towards contemporary social pertinence which involves certain ironies in the comparison. For while the course of the cultural decline traced by Leavis is roughly congruent with Eliot's version, the ideological traditions to which she appeals as positive factors are Eliot's (and indeed Arnold's) declared enemies: the Puritanism of Bunyan, the 'solid unromantic bourgeois interests' [*FRP*, 91] of Defoe, the civilizing work of Methodism. Although they join hands over many issues, Leavis has the advantage here in a more considered sense of social solidarity and responsibility inherited in part from the Puritan tradition.

The virtues of cultural coherence which Leavis holds up for contrast with twentieth-century disintegration are, in the first place, those of the Elizabethan age. Shakespeare, as F. R. Leavis had pointed out in *Mass Civilisation and Minority Culture,* entertained an audience embracing all social classes. Instead of being separately catered for by the cinema and the best-seller, 'the masses were receiving their amusement from above . . . They had to take the same amusements as their betters . . . even if they would not have understood the finer passages. Happily,' Leavis adds, 'they had no choice . . .' [*FRP*, 78]. As illiterates, the masses had no publishers pandering to them; as compensation for illiteracy, they possessed an enormously rich heritage of proverbial wisdom, music, ballads, and folk-history. They had a lore, while today's literate masses have only information, 'a kind of knowledge not rooted in the soil . . .' [*FRP*, 79]. There was as yet no distinction between journalism and literature; authors did their own advertising, conversation was still an art, and the reading public was still small enough to be 'a genuine community, as yet unspoilt by the traffic in literature' [*FRP*, 85].

The next phase in English taste which Leavis discusses is that represented by the four most widely read works for a century and more: the Authorised Version of the Bible, *Paradise Lost, The Pilgrim's Progress,* and *Robinson Crusoe.* This was the Puritan tradition, a tradition suddenly broken towards the end of the nineteenth century. For Leavis, this literature bears witness to a much finer and more satisfying quality of life among the common people at the end of the seventeenth century and among 'the shopkeeper class at the beginning of the eighteenth century'[7] than that enjoyed by their modern descendants. Again, it is a definite rootedness which sustains Bunyan: 'Bunyan's vigour derives from the soil' [*FRP*, 88], and his moral sense from the shrewdness of the English peasantry. Through him we can make contact with a genuine folk culture. 'Thus the characteristic effect of reading a passage of Bunyan is a stirring of the blood — the Biblical phrases and cadences evoke overtones, and the peculiarly thrilling quality of the prose is due to this technique which enables a precise particular occasion to draw on the accumulated religious associations of a race.' [*FRP*, 89–90.]

The solid bourgeois values of Defoe were perhaps crude, but he too drew upon, and appealed to, a public which did not want or need to pry into its own feelings or crave cheap sentiment. This Puritan tradition provided an improving course of good reading to compensate for lack of formal education in the lower classes. Upon this basis the polite journalism of the *Tatler* and the *Spectator* in the eighteenth century worked successfully to refine and polish the growing bourgeois readership, to bring it into a world of good breeding yet without losing touch with contemporary speech. Addison and Steele perfected a cultural consensus, a common vocabulary to combine the upper- and middle-class readerships into one: a reading public which in turn made possible the development of the English novel from Richardson to Austen. The novelists of the late eighteenth century were as rooted as Defoe had been in common speech and a common view of life. The seeds of decline, though, were sown in the same period, as newspapers greatly increased their circulations and Richardson discovered a new sentimental formula for the novel, which was to create an undesirable 'taste for novel-reading as distinct from a taste for literature' [*FRP*, 112] — a trend which circulating libraries rushed to exploit. Still, the best novelists set a high tone of decorum far removed from the emotional crudities and vulgarities of Dickens or of Charlotte Brontë; and the rate of decline was cushioned still further by 'the important fact that a fair proportion of the population could not read: it received its education through hand and eye and word of mouth and did not complicate matters by creating a separate semi-literate public to interfere with the book market . . . the rate of absorption of the lowest class into the middle class was slow enough to prevent any lowering of standards.' [*FRP*, 122.] A further restraint upon the malign influence of these raw recruits was the exorbitant price of novels, until the advent of serialization. And at the other end of the social scale, the culture of the educated classes was still thriving, firmly based upon family life: small nuclei of the educated still gathered in their homes for music and discussion. 'That is to say, the governing class was cultivated.' [*FRP*, 125.]

It was with the growth of periodical fiction that a much larger readership began to find novels within its economic means, and that the floodgates of bad taste were opened wide. The new sensibility is represented above all by Dickens, with his childish outlook on life, his 'crude emotional exercises' [*FRP*, 129]. The trend was accelerated with the arrival of mass-circulation newspapers for a similar public, used by powerful proprietors like Northcliffe 'to mobilize the people to outvote the minority, who had hitherto set the standard of taste without any serious challenge' [*FRP*, 151]. Leavis quotes the words of Sir Egerton Brydges to summarize the effects of these developments: '"Formerly, no doubt, the mob had a lower class of books than at present, but then they did not set them up for the best." It is, above all, the collapse of authority that marks the [modern] reading public . . .' [*FRP*, 152].

At least Brydges and even Arnold felt that their word could still command some deference from below, that they had an audience which would listen to their warnings. Now, Leavis argues, Arnold's 'anarchy' has become fact, and the cultural situation conforms to the fears expressed by Edmund Gosse in a passage already quoted by F. R. Leavis in *Mass Civilisation and Minority Culture*:

One danger which I have long foreseen from the spread of the democratic sentiment, is that of the traditions of literary taste, the canons of literature, being reversed with success by a popular vote. Up to the present time, in all parts of the world, the masses of uneducated or semi-educated persons, who form the vast majority of readers, though they cannot and do not appreciate the classics of their race, have been content to acknowledge their traditional supremacy. Of late there have seemed to me to be certain signs, especially in America, of a revolt of the mob against our literary masters. . . . If literature is to be judged by a plebiscite and if the plebs recognizes its power, it will certainly by degrees cease to support reputations which give it no pleasure and which it cannot comprehend. The revolution against taste, once begun, will land us in irreparable chaos.[8]

The problem, for Leavis as for those whom she quotes, is less the lowering of taste in itself than the undermining of standards which this implies when combined with the threat of democracy. It is in the counter-revolutionary struggle for taste that her anthropological or sociological method is brought into action.

Despite the emphasis given, in Leavis's history of cultural decline, to the influence of the semi-literate mass, when she comes to list the factors behind the lowering of standards she gives first place to a transformation of the governing class brought about by the public school system in particular. This replaced the old eccentric scholar and gentleman with the conformist sportsman: 'Altogether the character of the governing and professional classes has radically altered. The people with power no longer represent intellectual authority and culture.' [*FRP*, 155.] It is this betrayal by the ruling class that is responsible for the lack of cultural deference among the masses and for the isolation of the minority from power. The collapse of taste is not a result merely of the growing strength of the masses, but a tendency common to all classes. The sociological category which Leavis uses to account for this, to replace the redundant category of class, and to separate the cultural sheep from the goats, is the 'herd'.

The use of the term 'herd' in *Fiction and the Reading Public* is not just a device to relieve the monotony of terms like 'mob' and 'mass', but the signature of a definite sociological theory derived from *Instincts of the Herd in Peace and War*, the work of one W. Trotter. This book was to be listed in the bibliography of F. R. Leavis and Denys Thompson's *Culture and Environment* immediately after *Fiction and the Reading Public,* with a special recommendation for use as essay material by scholarship candidates. The theory which it expounds is a crudely biological social psychology in which different human nations are organized according to varying gregarious patterns observable in the animal kingdom: thus the Germans are like wolves and the English like bees. Within each national 'herd' there are subsidiary 'herds' or classes, whose competition impedes national unity. Human beings are classifiable into two great mental 'types': the stable, 'resistive', or conformist type who are extremely herd-like and deaf to new ideas or fresh experience, and the unstable, sensitive type, who are receptive to new experience and ideas. Ruling classes are invariably made up of the resistive type, while the sensitive are an outcast minority. In an age which threatens great peril to civilization, such a state of affairs cannot be allowed to go on: 'Society can never be safe until the direction of it is entrusted only to those who possess high capacity rigorously trained and acute sensitiveness to experience and to feeling.'[9] None of the classes or sub-herds in society is capable of putting reason before its own herd-prejudice. Only through the slow elevation of the standard of consciousness, into 'a kind of freemasonry and syndicalism of the intellect',[10] will reason begin to govern society, breaking down the sub-sidiary herds into the true herd of the nation and mobilizing our 'deeply buried combined national impulses'.[11]

This concept of the herd is much more versatile than 'mob', since it enables Leavis to merge together the cultural attributes of the masses with those of the governing classes. As F. R. Leavis had insisted, there was no essential cultural difference between the classes, and as Q. D. Leavis spells out in *Fiction and the Reading Public,* the conformity imposed by the public school and by the cinema have one and the same root. Resistance comes only from the sensitive minority. It is not that fellow-feeling or conformity with social conventions is destructive; on the contrary, the sensitive minority need their own solidarity, their own consensus of values and assumptions, based on the tiny reading public of the past, to set against the values of the herd:

Throughout Chapter 6 numerous references were made to the formative force of society, while in Chapter 7 an apparently identical force described as the herd is alleged to have overthrown the work of the previous ages. 'Society' was to be interpreted in the eighteenth-century sense in which, like 'the world', it meant a select, cultured element of the community that set the standards of behaviour and judgement, in direct opposition to the common people. Thus the highest definition of man was that of a social animal: the gregarious instinct he shares with sheep and wolves. . . . If one accepts the argument [of *Mass Civilisation and Minority Culture,* that fine living depends upon a tiny discerning minority], then it becomes evident that the individual has a better chance of obtaining access to the fullest (because finest) life in a community dominated by 'society' than in one protesting the superiority of the herd. [*FRP*, 162–3.]

In this definition of terms, the distinction between healthy gregarious instincts shared with animals and the threatening collectivity of the herd is very close to Trotter's conceptions, up to and including the favoured domination of society by a select cultured minority. The borrowing of the 'herd' model has important consequences for Leavis's anthropological or sociological method which, it becomes apparent in *Fiction and the Reading Public,* is hardly sociological at all, but essentially psychological. Trotter had argued that since so much of human behaviour is based upon instinct, 'sociology, therefore, is obviously but another name for psychology, in the widest sense'.[12] Leavis proceeds upon the same assumption.

While Leavis had argued in the Introduction to *Fiction and the Reading Public* that popular fiction was a proper object of study for anthropology rather than for literary criticism, when she comes to discuss the qualities of nineteenth-century novels, she decides that 'the bestseller becomes less a case for the literary critic than for the psychologist' [*FRP,* 135], citing as evidence a list of compound words beginning in 'sex-' taken from a popular novel. This strain of Leavis's work shows the influence of puritanism (at least in the loose sense of the word) in its ugliest light. She discusses the reading of bestsellers as a form of 'the drug habit',[13] refers at one point to 'the lascivious syncopated rhythms of twentieth-century song-and-dance records' [*FRP,* 83], describes the sentimental morality of nineteenth-century novels and Hollywood movies as 'largely masturbatory' [*FRP,* 136], and claims that romantic bestsellers promote 'a habit of fantasying [which] will lead to maladjustment in actual life' [*FRP,* 55]. Richards's pupil has taken his concern for mental health to new limits. As with him, the snap psychological judgement of personalities is not confined to any separable area of 'social comment' but appears at the heart of her literary-critical choices. Particularly when dealing with Dickens and Charlotte Brontë, Leavis brings in accusations of immaturity, day-dreaming, and retarded emotional development almost to the exclusion of literary (let alone psychoanalytic) evidence. Just as F. R.

Leavis had located Lawrence's value 'in his living person',
Q. D. Leavis makes this personal approach into a general
method for criticism of the novel, arguing for a kind of
analysis which dispenses with technical abstractions like plot,
character, theme, or setting.

. . . a discussion of the mechanics of successful novels (except for
professional novelists) is pointless and profitless. The essential technique
in an art that works by using words is the way in which words are used,
and a method is only justified by the use that is made of it; a bad novel
is ultimately seen to fail not because of its method but owing to a fatal
inferiority in the author's make-up. [*FRP*, 186.]

Quoting Henry James's assertion that the quality of a work
of art is the quality of the mind of the producer, Leavis
commits herself to a heavily psychologistic method of
criticism, in line with her view of society. The literary-critical
and the sociological methods indeed merge together, as for
example in a discussion of Joyce's character Gerty MacDowell,
who is cited as an 'invaluable reference' in understanding the
pernicious effects of stock responses in real life:

. . . for Gerty MacDowell every situation has a prescribed attitude
provided by memories of slightly similar situations in cheap fiction, she
thinks in terms of clichés drawn from the same source, and is completely
out of touch with reality. Such a life is not only crude, impoverished,
and narrow, it is dangerous. And it is typical of the level at which the
emotional life of the generality is now conducted. [*FRP*, 195-6.]

Such a method of making confident sociological statements
about the attitudes of the 'generality' upon purely literary
evidence is itself crude, narrow, and dangerous. Richards too
had somehow discovered that 'in fact, the idle hours of most
lives are filled with reveries that are simply bad private
poetry' [*PC*, 320]. And it is surely remarkable that Leavis
should pronounce upon the inner thoughts of millions on
the sole basis of − slightly similar situations in expensive
fiction. From this passage it seems that Gerty is not the only
one who is mistaking fiction for reality. Leavis's tendency to
collapse fiction and reality together, to regard literature as
a transparent register of the thought or sensibility of its
producers and consumers, has further consequences for her
sociological method. Thus, for example, she can claim (and

this is at the heart of her historical argument) that 'the spectator of Elizabethan drama, though he might not be able to follow the "thought" minutely in the great tragedies, was getting his amusement from the mind and sensibility that produced those passages, from an artist and not from one of his own class. There was then no such complete separation as we have just seen to exist between the life of the cultivated and the life of the generality.' [FRP, 209.] When Leavis says 'life' she is referring in fact only to literature. The differences of class which she mentions are dissolved in the same breath by the fact that people of all classes watched the same plays — and even that, if one includes the tradition of courtly masque, is not strictly true. The sociological method breaks down into a model built upon purely literary impressions.

It is not that sociology is entirely smothered by this literary/psychological view, however. A skeleton sociology is still discernible in *Fiction and the Reading Public,* though ultimately founded upon literary and psychological assumptions: there is the herd, embracing all classes, on the one hand and the discerning minority (or 'society') on the other, and in addition a third social force which takes on a particular importance within such a model — the 'middlemen' or fixers of herd mentality. Since the essential characteristic of a herd is its passivity, its suggestibility, it follows that 'the majority has its mind made up for it', that the bulk of the reading public 'has no means of knowing what it really thinks and feels' [FRP, 33, 195]. The initiative rests entirely with an all-powerful minority of opinion-formers: 'The extent to which the attitude approved by the herd is fixed by such agencies for imposing conformity as the Public Schools, advertising, and the Press, cannot be overestimated'. [FRP, 157.] F. R. Leavis had insisted that power in society lies with the machine. Q. D. Leavis adds that this power is wielded, as if on the machine's behalf, by the media and the educational system. To prevent the abuse of such power it will be necessary to establish alongside the legislative and executive powers of machine and media an independent judiciary — literary criticism — which can scrutinize and challenge their control of the herd.

. . . they have assumed such a monstrous impersonality that individual effort towards controlling or checking them seems ridiculously futile. This is probably the most terrifying feature of our civilisation. If there is to be any hope, it must lie in conscious and directed effort. All that can be done, it must be realized, must take the form of resistance by an armed and active minority. [*FRP*, 213.]

The minority, again, is the self-defining readership which is to take Leavis's case as a reminder of the obvious rather than as an argument — those who recognize themselves when Leavis remarks that literature's 'importance can be assumed to need no demonstrating to any reader of this essay' [*FRP*, 169]. This minority is to be 'armed' with an independent press and an all-round critical journal, and has two fields of action: research along the lines of *Fiction and the Reading Public,* and the sending of a picked band of missionaries into the educational system to organize resistance to advertisements, movies, and bestsellers.

Given the overwhelmingly depressing contemporary background which Leavis paints in her conclusion to *Fiction and the Reading Public,* it is worth scrutinizing the one concrete example she gives of the effectiveness of missionary resistance work. The example given is, bafflingly enough, British Honduras. 'Here, I am informed, we have a community which in deliberately setting out to resist American influence is actually preserving a traditional way of life.' [*FRP*, 214.] The very remoteness of the place is enough to cast doubts upon the relevance of such resistance to Britain; but the facts (as distinct from literature or rumour) about the colony are even more crushingly eloquent against Leavis's projections. Far from resisting American influence, the very boundaries of British Honduras were fixed by a treaty between Britain and the USA (the Dallas–Clarendon Pact of 1856); trade was overwhelmingly dependent on the United States to the extent that exports collapsed from over $5m. to just over $1m. in the three years following the Wall Street crash of 1929; and of the two products upon which the unbalanced economy was almost entirely dependent (mahogany and chicle), the chicle resin was used — before being wiped out by synthetic substitutes — as the base for that most American of all consumer goods, chewing-gum.[14] One (non-Marxist) economist who has studied the colony describes it as 'a classic of colonial exploitation, of taking away and not giving back'.[15] Agriculture, the traditional 'roots' sought by the Leavises, was terribly neglected at the expense of forestry for the US market, to the extent that the colony had to import a third of its food. As for the preservation of a 'traditional way of life', it is difficult to see what is being referred to. The original purpose of the settlement, for nearly two centuries had been logwood-cutting, but this trade had just been wiped out by progressive deforestation without proper replanting and by the crash of markets during the First World War. The Maya Indians were abandoning their agriculture for the American chicle industry. Not a single school had been founded by the colonial government, and illiteracy was very widespread among the predominantly black and Indian population. Possibly Leavis is referring to the comical transplantation of English sporting culture to the colony in the 1920s, represented by the starting of an annual rowing regatta in 1927, together with the installation of a golf course and cricket ground. These projects had been funded by a legacy from the eccentric Barón Bliss, who had stipulated that none of it was to be spent on schools. If it is a minority culture one is looking for, a small colony of former slaves ruled and owned by a handful of whites and plagued by poverty and illiteracy is of course a fairly safe place to start looking, and to this extent Leavis was quite consistent in pointing to such an unlikely-sounding source of hope. Organizing a 'resistance' in Britain itself, though, was a larger problem altogether, requiring a programme of re-education and immunization against advertisements and popular entertainment. This was the aim of a separate work, *Culture and Environment*, by F. R. Leavis and Denys Thompson.

iii. *The power of advertisements*

Years later, F. R. Leavis was to insist that the enormous influence of *Culture and Environment* was in turn derived from its debt to Q. D. Leavis's work in *Fiction and the*

Reading Public. Using her research as a basis, Leavis had been
able to write *Culture and Environment* in a week, with Denys
Thompson's advice on presentation for schools.[16] Nevertheless,
the book's merits are of a very different kind, concentrating
the work of both Leavises into the educational programme
which they had promised. Almost a third of the text is made
up of recommended essay or project questions, and a booklist
for classroom use, complete with prices, is provided. This
work is Leavis's contribution to 'practical' criticism in the
sense of attempting to change the world; his alternative to
political commitment; his projection into the school curriculum of the *Scrutiny* critique of contemporary civilization.
It is here, too, that Leavis's ideas of the 'organic community' and of the importance of advertising find their most
cogent expression.

Leavis's first address is to the morale of the teaching
profession — his chosen agency of cultural resistance. To
quicken the teachers of English with a sense of their importance for the future, Leavis includes them in Richards's
evolutionary vanguard: 'The instinct towards health — the
instinct of self-preservation — that we must believe to be in
the human spirit will take effect through them or not at all.
In a world of depressed and cynical aimlessness there is for
them work that, aiming at considerable and considered ends,
will yield enough in immediate effect to make whole-hearted
devotion possible.' [*CE*, 4.] The new inspiration which Leavis
offers is the possibility of 'the training of critical awareness'
(the book's subtitle) sufficient to immunize pupils against the
whole weight of civilization. 'The aim of education should
be', Leavis insists, no less than 'to give command of the art of
living' [*CE*, 107]. This aim is ambitious enough in itself, but
becomes more arduous still when the antagonistic forces of
death are taken into account. For there is a more alluring
'education' on offer from the newspapers and advertisements,
against which the art of living must be defended. 'It is plain
a modern education worthy of the name must be largely an
education *against* the environment of which this passage
[a plan for a beer advertising campaign] is representative.'
[*CE*, 106.] In this way teachers of English are given a new
vision of their job as a daily struggle as inspiring as the

class war is to the Marxists, and, for Leavis, more practical.

Education's most important antagonist by far in *Culture and Environment* is the advertising business, both in its own right and in its degrading influence upon the press. There had been strong hints already in *Mass Civilisation and Minority Culture* that Leavis saw advertising as a major factor in cultural decline, and again in some of his early reviews. This outburst appears in 'John Dos Passos': 'can a hundred D. H. Lawrences preserve even the idea of emotional sincerity against the unremitting, pervasive, masturbatory manipulations of "scientific" Publicity, and, what is the same thing, commercially supplied popular art?' [*FC*, 105.] The occasion for this is Dos Passos's character J. Ward Moorhouse, a cynical advertising executive whom Leavis takes as representative of the power which holds society together, and as Dos Passos's only successful character portrait. For F. R. Leavis he serves a representative function similar to that of Gerty MacDowell for Q. D. Leavis. The invisible 'link' between these two — Joyce's Leopold Bloom, who is both advertising clerk and reader of pornographic fiction — is a case worth considering as a clue to the Leavises' noted silence on Joyce's enormous achievement as a novelist. For Joyce's adoption of such a figure as (however ironically) a sympathetic hero indicates a certain artistic humility and generosity which the Leavises could not rise to meet. In their sociology of the 'herd' a Leopold Bloom could find no place, squeezed out by the MacDowells and the Moorhouses into an impossible blind spot. The principle which would exclude him, and which Leavis expounds insistently in *Culture and Environment*, is that of the perfect automatic communion of advertiser and herd achieved by the scientific exactitude of marketing psychology:

The advertising expert, guided by years of carefully recorded and tabulated experiment (there are Schools of Advertising), sets 'scientifically' to work to get a given reaction out of the public he has in mind, concerning the relevant characteristics of which he has something like exact knowledge. He devises his appeal in the confidence that the average member of the public will respond like an automaton. [*CE*, 12.]

For Leavis, advertising is so powerful that it can create

personalities, or turn them into automata. From the evidence of advertising he suggests that there is a connection 'between standardization of commodities and standardization of persons' [*CE*, 32]. Accordingly, the critique of advertising reverts, as with Richards and Q. D. Leavis, to criticism of the imputed mentality of its consumers. The essay questions to be set on particular advertisements take the form, 'What kind of persons can you imagine as responding to such an appeal as this last?' – or (on a herd-like character in an advertisement): 'What do you think his attitude would be towards us? How would he behave in situations where mob passion runs high?' [*CE*, 51, 17.] The inclusion here of the 'us' who are threatened by mob passion is characteristic of all Leavis's writing, attempting to conjure against the *bonhomie* of the advertiser's appeal an equal and opposite solidarity of the discriminating, an intimacy insinuated in such questions as 'Why do we wince at the mentality that uses this [American] idiom?' [*CE*, 121]. 'We' are a composite mentality whose assumptions are to be as unanimous and as mutually reinforcing as those of the unitary advertiser/herd (and the bourgeois/Marxist) mentality.

To understand fully the basis of the Leavises' criticism, the literary historian has to be prepared to follow them into unfamiliar territory. As Q. D. Leavis had had to engage in sociology in order to trace certain social roots for the literary minority, so F. R. Leavis, in accounting for the enormous power which he attributes to advertising, is obliged to try his hand at economics. To 'wrest from the economist' the idea of the quality of life, it is necessary to organize a foray into the economist's field. The exercise is entirely to Leavis's discredit, revealing only the pitfalls of the 'literary' method as a starting-point for the understanding of society. The first and most extensively repeated error is the exclusive use of statements by advertising agencies as proof of the unlimited power and expertise of their profession. It seems never to occur to Leavis that advertising is a trade which thrives by advertising *itself*. He thus falls for the biggest advertising 'con' of them all, taking as good coin agency boasts designed to attract investment in publicity whose value has never been conclusively proved. While appearing to challenge and to scrutinize the output of the advertising business with the utmost vigilance, Leavis's literary/psychological method is in fact grounded on a fathomless gullibility, since in leaping from the advertisers' boasts to the ugly 'sensibility' behind them, it never stops to enquire into the motives for these quoted claims. The literary method seems to be able to inspect a text exhaustively for signs of unhealthy mentality, but since this mentality is seen as the 'truth' of the text, its real factual content goes unchecked. If the text's statements are fraudulent, the critic can only reproduce the fraud in inverted form. The consequence is a massively inflated estimate of the importance of advertising in the modern economy. Leavis twice quotes without qualification the view of the *Criterion*: 'The material prosperity of modern civilization depends upon inducing people to buy what they do not want, and to want what they should not buy.'[17] While it is quite conceivable that the readership of the *Criterion* was unaware of the existence of rather large numbers of people who could not *afford* to buy what they did not want (and for whom therefore advertising's effect was severely limited), Leavis at least had indicated his recognition of this economic and social fact. He chooses, though, to ignore, indeed invert it, as if the majority of the herd were in any economic position to respond like automata — as he claims — to the enticements of advertising. His faith in the power of the Word to make people spend more money blinds him to the real economic limits to consumer demand, which can only be extended by more worldly economic measures like increases in credit and wages. The statistical facts of advertising's economic effects (as opposed to the shared claims of advertiser and critic) are overwhelmingly against Leavis on this point. The painstaking research of Neil H. Borden in the homeland of advertising — the United States — concludes that the effect of advertising on overall demand is negligible.[18] Research in Canada has found that consumers develop their own 'self-defense mechanism' against the claims of advertisements,[19] apparently without the aid of English-teachers; and an exhaustive study of advertising statistics in western Europe finds that even in advertising-intensive markets it is *price* which remains a more important determinant of market shares.[20] The findings of independent and rigorous research, in short, flatly contradict the assertions of *Culture and Environment*.

Certainly, Leavis should not have been expected to restrict his critique of advertising to its cultural effects, its occasional bad taste, or insulting assumptions. If he had limited his economic assertions, though, to the duplication and waste of resources involved in advertising, and to its financial influence upon the Press, he would have been on quite firm factual ground. But instead, Leavis proceeds to compound his errors to their logical limit, asserting that the modern capitalist economy cannot even function without the advertiser: 'Immediately it is in place to note the relation between mass-production and the abuses of advertising. A mass-production plant can be worked profitably only if it is worked to its full capacity, and only if its full output is absorbed by the market.' [*CE*, 30.] After making this assertion, Leavis immediately goes on to state that full output for needs rather than for artificially created demand can only be achieved in a planned economy, whereas capitalist factories are run competitively and for private profit — with a reminder here to the teacher to be 'discreet' about following this argument through. Again, the apparent proximity to Marxism only highlights a real remoteness, for the implication is that advertising is the sole guarantee of profitability: Leavis's adoption of the teaching profession rather than the class war as a 'revolutionary' agency begins to make sense in the light of this argument. The prospect opens up, for the 'armed minority', of a bizarre inverted bolshevism of the market-place, undermining the capitalist economy not from the point of production but at the point of consumption. Logically, if market saturation by means of advertising is a necessary condition for private profit, and enough people are immunized against the temptations of advertisements, the entire capitalist system should come crashing down like the walls of Jericho.

Fortunately for capitalism, Leavis's assertion that mass production can be run profitably only at full capacity is a patent falsehood. To call it a howler would be a charitable description, since it is not just a lack of theoretical training which is involved. Had Leavis bothered to read any Marx, or even a school textbook of economics, he might have seen the error; but he did not even need to delve into such abstract and dogmatic sources, since the evidence of contemporary 'concrete experience' was abundantly available in the early 1930s in the form of an acute economic depression — precisely the ideal conditions in which to observe the fact that whole industries can and do operate profitably below full capacity, given a variety of factors among which advertising scarcely deserves a mention. Wild though Leavis's inaccuracies are, they serve their purpose admirably: to enhance enormously the social importance of literary critics and of English teachers, beyond even the heights at which Arnold and Richards had pitched their claims. The sociological and economic assumptions of *Culture and Environment* reduce the choice upon which civilization's future depends to a battle between two market consumption regulators (to use the — for Leavis — unspeakable language of the economist), advertising and literary criticism: a battle for the hearts and minds of the passive and suggestible multitude. The weapon common to, and massively overestimated by, each side is the limitless power of the Word.[21]

The struggle would be stalemated, though, without a tradition upon whose strength the literary minority could draw. In its literary or pamphleteering form, the tradition is recognizable from the reading list at the back of *Culture and Environment,* which includes Bunyan, Cobbett, Arnold, Richards, and Q. D. Leavis. But it involves much more: an entire 'organic community' of the old English village, of the natural rhythms of the soil, of handicrafts and agricultural self-sufficiency, all evoked in long quotations from George Bourne's *Change in the Village* (1912) and *The Wheelwright's Shop* (1923). In this lost world, work had a satisfying purpose, a more than merely economic function; workers in handicraft trades were 'men' and not 'hands' — 'Just as their master was not concerned merely for his profits, so they were not concerned merely for their wages.' [*CE,* 75.] Of this organic community, Leavis writes: 'Its destruction (in the West) is the most important fact of recent history' [*CE,* 87]. But he rejects any possibility of recovering this lost world, of scrapping the machine.

The point to be grasped here is not that Leavis's view of

the 'organic community' is nostalgically idealized (which it is) or that he is trying to reinstate it (which he is not), but the use to which it is put for his own time. It is the *memory* of the old order which is recoverable and valuable today. And what was most regrettable in the loss, most urgently in need of renewal, was the sense of uncomplicated transparency in social relations implicit in Bourne's formula 'the coherent and self-explanatory village life', which Leavis quotes as 'something of the first importance' [*CE*, 74]. For Leavis, the self-explanatory nature of the village community was its most attractive quality, of such importance that it underpins his whole literary-critical project. In an article on modern criticism in 1932 he had written: 'The dissolution of the traditions, social, religious, moral and intellectual, has left us without that basis of things taken for granted which is necessary to a healthy culture. A serious literary critic nowadays cannot confine himself to literature.'[22] In ranging further afield, it is the taken-for-granted, the 'implicit' that the Leavises take with them from literary criticism into sociology, history, economics, or politics. The unproblematic 'community' or consensus that they understand to be built into the act of literary criticism itself, becomes the centre of social renewal. It is in this sense that Leavis states that 'literary education . . . is to a great extent a substitute' [*CE*, 1] for the organic community. In the work of the Leavises, Arnold's hopes for literature as a replacement for religion and philosophy are both renewed and extended. Literature becomes not so much a substitute religion, more a substitute way of life.

NOTES

1. *The Use of Poetry and the Use of Criticism*, 108.
2. F. R. Leavis, *New Bearings in English Poetry* (1932; 2nd edn. 1950), 13.
3. *FC*, 10. In the light of this position, Lionel Trilling's view of Leavis, that 'Class is for him a cultural entity' (*Beyond Culture* (1966), 171), is seriously misconceived. An even worse error was made by Arthur Calder Marshall in the *New Statesman*, where he accused Leavis of advocating 'Communist tyranny'. Leavis hastened to point out his

anti-Marxist credentials: see his *Letters in Criticism*, ed. J. Tasker (1974), 28.

4. *FC*, 184–5 ('Restatement for Critics').

5. Francis Mulhern, in *The Moment of 'Scrutiny'*, locates Leavis's fundamental divergence from Marxism in his conceptual conflation of the social relations of production and the forces of production as economic determinants, abolishing the contradictions which Marxism traces between the two terms (pp. 72–6). Though this account tends to minimize other fundamental differences, in particular Leavis's insistence on the essential passivity of the masses, the consequences of Leavis's view of economic determination are well presented in Mulhern's conclusion: the effective dissolution and repression of politics (p. 330).

6. *FC*, 110 ('John Dos Passos'); cf. *FC*, 5, 94.

7. *FRP*, 87. The 'roots' sought here are not fanciful. Q. D. Leavis's father was a draper; F. R. Leavis's a music-shop keeper.

8. *FRP*, 154, from Gosse's *Questions at Issue* (1893), 110–11.

9. W. Trotter, *Instincts of the Herd in Peace and War* (1916; 2nd edn., 1919), 240. It is probably this work to which Bronislaw Malinowski refers in an appendix to Ogden and Richards's *The Meaning of Meaning*, where he writes that 'the term Herd-instinct has been misused in a recent sociological work' (p. 314n).

10. Ibid., 259.

11. Ibid., 206.

12. Trotter, *Instincts*, 11.

13. *FRP*, 22. Cf. Arnold on believers in religious miracles: 'They are like people who have fed their minds on novels or their stomachs on opium; the reality of things is flat and insipid to them, although it is in truth far grander than the phantasmagorical world of novels and of opium. But it is long before the novel-reader or the opium-eater can rid himself of his bad habits, and brace his nerves, and recover the tone of his mind enough to perceive it.' [*CPW* vi. 379 (*Literature and Dogma*).]

14. See Stephen L. Caiger, *British Honduras Past and Present* (1951).

15. N. S. Carey Jones, *The Pattern of a Dependent Economy* (Cambridge 1953), 18.

16. Leavis, *Letters in Criticism*, 58.

17. *CE*, 26, from *Criterion*, viii (1928), 189.

18. Neil H. Borden, *The Economic Effects of Advertising* (Chicago, 1947), 433.

19. O. J. Firestone, *The Economic Implications of Advertising* (Toronto, 1967), 181.

20. Jean Jacques Lambin, *Advertising, Competition and Market Conduct in Oligopoly over Time* (Amsterdam, 1976), 161–2.

21. In fairness to Denys Thompson, it should be recorded that his later work on advertising, *Voice of Civilisation* (1943), disowns the idea that advertising is central to the capitalist system, and sees it only as a symptom of that system. Even here, though, Thompson maintains one of the major fallacies of *Culture and Environment* when he states that 'people living in masses are more open to suggestion than those living in

small, organic groups' (p. 72). The entire history of religious 'suggestion'
ignored here shows exactly the opposite tendency.

22. F. R. Leavis, 'This Age in Literary Criticism', *Bookman*, lxxxiii
(1932), 9.

8.   A COMMON PURSUIT: SOME CONCLUSIONS

> 'Do you think so?' said Gwendolen, with a little
> surprise. 'I should have thought you cared most about
> ideas, knowledge, wisdom, and all that.'
>
> George Eliot, *Daniel Deronda*

By 1932, the year of *Scrutiny*'s launch, English literary
criticism had undergone, as F. R. Leavis noted at the time,
a considerable revival at the hands of Eliot and Richards
— a revival sharpened by the sense of isolation of a self-
conscious literary minority. Frustrated by the obstacles
to the kind of social influence due to them according to the
projections made by Arnold, critics and teachers of English
literature extended Arnold's claims for the centrality of
poetry in modern life, at the same time elaborating Arnold's
implicit analysis of the Philistine society in which literary
culture had to do its work. The revival was sustained and to
a certain degree consolidated by the culmination of an
extended campaign for the study of English literature at
schools and universities. The long march of the Philistines
from the dissenting academies and Mechanics Institutes to
Oxford and Cambridge, spreading into the school curriculum
and into wider areas of higher education, had achieved its
basic institutional goals. The English Association, with its
aim of 'establishing the recognition of English as an essential
element of the national education' largely secured, began to
see a dropping off in membership in the early 1930s as the
effects of the Newbolt Report removed the need for pressure-
group activity. What this educational movement left behind
it as it began to ebb was a particularly concentrated deposit.
Neither at London nor at the younger colleges was there the
strength and the personnel to carry the loftiest ambitions of
the movement as expressed in the Newbolt Report; at Oxford
under the reign of Raleigh and Gordon such high hopes for
literary study as a renewal of civilization were actively
dampened; only at Cambridge, where the English School had

been exposed at birth to the inflammatory influences of born-again wartime Englishness, could the combined advantages of ancient precedence and a large demobilized student intake be harnessed to establish English as a truly 'central' subject. With no equivalent of 'Greats' — the predominant humane discipline at Oxford — Cambridge had the basis upon which English Literature could become that kind of 'subject' in both senses: an organized academic discipline, and at the same time a more or less conscious agent in the world. The innovations of Eliot and Richards enabled what had been suspected of being a soft option, a frivolous subject, to adopt that appearance of strenuousness and difficulty proper to a serious branch of study; and the Cambridge English tripos was the first to embody this rigour in an examination designed to be testing in a more extended sense than usual. As Q. D. Leavis described *Wuthering Heights* as a 'total-response novel', so the Cambridge practical criticism paper was to be a total-response examination, challenging the entire range and resources of a candidate's unaided sensibility. This was the centre-piece of English studies at Cambridge; to complement it, there evolved under the guidance of Willey and Tillyard a made-to-measure version of history draped around it as 'background' material.

The remarkably inventive critical production of T. S. Eliot in the years 1917-24 had been worked up into an acceptable technique by I. A. Richards's work of 1924-9, codified in examinations, and brilliantly exercised by Richards's student William Empson in *Seven Types of Ambiguity* (1930). But as the achievement accumulated upon successive shoulders, the initiators peeled away: Eliot to a diminishing circle of clerical reactionaries, and to a position where he ridiculed the teaching of English literature in England as an absurdity; Richards via Coleridge and Confucius to the elaboration of Basic English as a medium of world communication and reconciliation; Empson, following Richards, to China. Those left to guard the Cambridge fort were the younger critics headed by the Leavises — rooted in the English tripos, but for all that marginal to the English School as an actual institution: for them, an outlaw status which placed them at the fringes and at the 'centre' of Cambridge at the same time.

As F. R. Leavis recalled in his 1963 'Retrospect' to *Scrutiny*:

Cambridge, then, figured for us civilisation's anti-Marxist recognition of its own nature and needs — recognition of that, the essential, which Marxian wisdom discredited, and the external and material drive of civilisation threatened, undoctrinally, to eliminate. It was our strength to be, in our consciousness of our effort, and actually, in the paradoxical and ironical way I have to record, representatives of that Cambridge. We *were* in fact that Cambridge. We felt it, and had more and more reason to feel it, and our confidence and courage came from that.[1]

Yet in the same article Leavis speaks of *Scrutiny* as a project inconceivable without the English tripos, surviving 'in spite of Cambridge' as 'an outlaw's enterprise'.[2] The status of this group seems oddly similar to that of Arnold's Scholar Gipsy, and with similar ambivalent advantages: capable of appealing to an ideal 'centre' of which they were the real periphery, and identified all the more with the ideal the more the real, and at times petty, academic politics of the English School excluded them. Partly because of this status as at the same time Cambridge and not-Cambridge (as Arnold had positioned himself to be both Oxford and not-Oxford), it was this team who were to realize the accumulated potential of the Arnoldian critical tradition as reworked by Eliot and Richards, and to organize it into a sustained and distinct movement under the name of *Scrutiny*. The combined powers of Eliot, Richards, and the Leavises were not confined in their influence to the immediate followers of *Scrutiny,* either, but spilled over in a number of directions: one side-effect, for example, was the dethronement of Bradleian 'character' in Shakespeare criticism, identified with the work of G. Wilson Knight and L. C. Knights in the early 1930s; and further afield, these influences were to feed into the development of a powerful critical orthodoxy in North America — the New Criticism.

In studying the evolution of ideas on literary criticism's social functions, the foregoing chapters have attempted to examine separately the major figures or groups of lesser figures behind the English critical renaissance of the 1920s and early 1930s associated with the term 'practical criticism'. It remains to summarize the common and roughly consistent ingredients of the critical outlook which these writers share —

a community and a consistency essential to their predominance in English criticism through the middle of the twentieth century. These common themes will be approached in this chapter in two groups: first, those concepts and practices which these critics tended to define themselves against negatively — science, theory, mass literacy, and the idea of history; secondly, after this negative definition, the positive characteristics of these critics' thought — their concept of order, and the racial, psychological, and social models which it encompassed, and their attitudes to literature and criticism as possible replacements for religion.

i. *Science*

Hostility towards science, and fear of its growth, is a constant and indeed constitutive element of the English literary-critical tradition from Arnold through to the Leavises. Arnold's projection of the immense future of poetry as a substitute for religion was made in the recognition that religion was too poorly placed to resist the encroachments of science, having attached itself too much to disputable facts. Poetry, on the other hand, could hold back the all-encompassing claims of the sciences more effectively, speaking to the 'whole man', whereas science, according to Arnold's scheme of independent faculties or 'sides' to human nature, could address itself only to a fraction of our personality. As was pointed out in Chapter 2, Arnold's scheme of 'intellectual deliverance' was devoted partly to denying science's claims to human relevance, and to reducing scientific method to the simple collection of facts, reserving all synthetic understanding for the humanist or belletrist. In conformity with his general principles of rounded human development, Arnold attempted to discourage the popular dissemination of scientific and sceptical thought on the grounds that this dissemination was premature and should await some future synthesis with religious thought. Until then a kind of intellectual quarantine should operate, according to which scientists should stick to their own departments and not challenge the dominant religious ideology through which people (and the Populace in particular) saw the world in which they lived. His elevation of poetry to the supreme position in human affairs was premised on the belief that science was 'incomplete'; most of Arnold's remarks upon science are attempts to ensure that science remain as incomplete as possible, that its fields of investigation and its disturbing conclusions be fenced off from the vulnerable areas of religion and ethics.

Despite the fact that the growth of scientific and technical instruction was an important factor in promoting English (rather than Greek or Latin) literature in the educational curriculum, little gratitude was shown by the beneficiaries of this development. The early advocates of English studies tended to echo Arnold's conception of science as a mere aggregation of 'dry' facts, mentally fissiparous in its effects and somehow inhuman, needing the complement of literature to humanize it. At its most far-fetched, this tendency appears in Walter Raleigh's passionate defence of poetry's blooming flowers of experience against science's dried weeds of abstraction, in which an equation is drawn between science and death, poetry and life (ironically, a far greater abstraction, this, than any scientist would dare venture). Much of the hostility towards philology, which aided the independence of English Literature as an academic discipline, was fuelled (as in Churton Collins's case) by such prejudices, in addition to the more prominent anti-German component in that campaign. Quiller-Couch, for example, frequently combined the two themes in his wartime lectures.

The post-war critical 'revolution' might appear to involve a casting-aside of these attitudes, in its concern for rigour and impersonality. Eliot's famous announcement of the 'impersonal' theory of art in 'Tradition and the Individual Talent' seems boldly to offer such a reversal: 'It is in this depersonalization that art may be said to approach the condition of science. I shall, therefore, invite you to consider, as a suggestive analogy, the action which takes place when a bit of finely filiated platinum is introduced into a chamber containing oxygen and sulphur dioxide.' [*SW*, 53.] Before this was reprinted in *The Sacred Wood*, readers of the *Egoist* had to wait three months before Eliot would solve the little mystery for them; and it seems clear that the analogy was introduced at this point largely for effect. In the second part of the article Eliot discloses that the platinum is an inert catalyst which triggers a reaction between the two other elements to produce sulphurous acid, while remaining itself unchanged by the reaction. In the same way, Eliot suggests, the artist is or should be neutral and unaffected by the feelings and emotions which are combined in the work of art. The analogy is vivid and more than surprising enough to enforce important reconsiderations in poetics. What is much more doubtful is the supposed connection (in the 'therefore') between this analogy and the foregoing suggestion of art's approximation to science. For the analogy is not between artist and scientist but between artist and platinum, and platinum (or even the concept of the catalyst) is not science, but simply one of its possible objects. Eliot is, in other words, perpetuating the mystique of science as a bewildering array of discrete objects, and exploiting this mystique for dazzling effect, rather than bringing about any rapprochement between the methods of art and those of science.

Eliot called Romanticism a short cut to the strangeness of life without the reality; but there is much in the pseudo-scientific trappings of his own anti-Romanticism which attempts similar short cuts to the authority, indeed the glamour, of science. This is evident in the arrogance of his statement on Aristotle, who 'had what is called the scientific mind — a mind which, as it is rarely found among scientists except in fragments, might better be called the intelligent mind' [SW, 13]. Eliot here is attempting to steal the name of science while giving nothing in return to its practitioners. Elsewhere in *The Sacred Wood*, he reverts to a more orthodox Arnoldian attitude when he discusses anthropology and psychology. 'I do not deny the very great value of all work by scientists in their own departments,' he writes, with the implicit stress on the qualification; but it is the task of the creative mind, rather than the scientific, to 'digest the heavy food of historical and scientific knowledge that we have eaten . . .' [SW, 75, 77]. As in Arnold's scheme, the task of synthesizing the fruits of specialist research is reserved for the artist. Not much later, Eliot was to abandon his scientific pose altogether, as the identification of science with other threats to tradition became more important to

him; as when he had discussed the heavy food of anthropological knowledge as the product of a time when 'social
emancipation crawled abroad' [*SW*, 75], he now identifies
science with the threat of democracy:

The man of letters, or the man of 'culture' at the present time is far too
easily impressed and overawed by scientific knowledge and ability; the
aristocracy of culture has abdicated before the demagogy of science. . . .
Democracy appears whenever the governors of the people lose their
conviction of their right to govern: the claims of the scientists are
fortified by the cowardice of men of letters.[3]

The 'scientific' flavour of Eliot's pronouncements in *The
Sacred Wood* was part of this war of nerves, in which he was
prepared boldly to appropriate the enemy's weapons — to
impress and to overawe — in much the same way as he raided
the works of past writers.

A similar contradiction appears with I. A. Richards,
between the language and aura of scientific rigour adopted
and the general tendency of his work to neutralize the human
consequences of scientific discovery. Richards's adoption of
the 'clinical manner' is not quite such a pose as Eliot's, since
he was more seriously aware of the methods of linguistics and
psychology, at least. Yet the same anxiety to cut the realm
of 'values' free from threatening scientific encroachments is
at work — all the more insistently for this greater acquaintance,
it appears. Behind the diagrams and the jargon, the central
message of Richards's work was not lost on his first students:
Basil Willey's recollection, cited in Chapter 6, was that
Richards offered a new vision of poetry as a stronghold
against scientific aggression. The portrayal of science as an
aggressor is a central feature of the account of threatened
cultural chaos in *Science and Poetry*. The growth of science
is one of the 'onslaughts of the last century' [*SAP*, 82] which
have forced our traditions back beyond their Hindenburg
line to the last defensive resort of poetry. But there is more
to Richards's view of science than this defensive stance; he
maintains that science can do no more than improve practical
control over nature, and not help us order our experience.
What we know and how we behave are two entirely separate
questions — as Richards and Ogden put it in *The Meaning*

of Meaning: 'It is not necessary to know what things are in order to take up fitting attitudes towards them . . .'.[4] Only the arts broaden the mind, only in art can we compare our experiences, and only through art can we organize ourselves; science, which might seem of service in these activities, is seen as an aggravation of the confusing variety of facts in modern life. Given these assertions, it is no surprise to find Richards referring, in a later work, to 'The Humanities, as the studies which make men human . . .',[5] as if that large majority of persons not connected with arts faculties were beasts or automata. Science, in full accordance with Arnold's formula, is relegated to the inhuman. The same approach is maintained by F. R. Leavis, who acknowledges the need for the vanguard of the race to be aware of science only in so far as it affects 'the sense of the human situation' — which to judge by the Leavises' writings means hardly at all: so persistently had science's importance been denigrated by their forbears that it goes almost completely unmentioned in the Leavises' early work.

ii. *Theory*

Closely connected with this common attitude to science is a constant and deeply held distaste for theory among these critics, and an accompanying tendency which amounts to a cult of raw experience. This is not just an important characteristic of their expository method, but a major element of their message: the need to displace from precedence these shadows and dreams and false shows of knowledge, as Arnold put it. The classical formula for this campaign is Arnold's 'Poetry is the reality, philosophy the illusion', but the same idea recurs throughout his work in different forms, from the rejection of 'depth-hunting' in his early letters to the later lecture whose advice was *'Don't think'*; from the refusal of definitions in favour of 'poise' to the claim that truth is found, not made or proved. We can reach truth, Arnold maintains, by looking hard and straight at the object, not by allowing our attention to be distracted by systems. Similar themes are repeated, almost lyrically, by Pater in almost all his writings, by Raleigh in his attack upon 'the way

of abstraction', and by Quiller-Couch, for whom the abandonment of definitions and theories in favour of personal tact
and judgement was a matter of pride. For Eliot, theory or
philosophy is to be excluded not so much from the critic's
method (which is to be based on comparison and analysis) as
from poetic works, though it is often implicit in his discussion
of this problem that his admiration for the artistic 'mind so
fine that no idea could violate it' extends beyond the limits
of literature as an exemplary quality of mind in general: a
'sensibility' in which systematic thought is sacrificed, transmuted, or digested beyond recognition. The constant attention
which Eliot devotes to the question of poetry and belief
reveals a major concern of his, to preserve an enclave within
culture where philosophy's writ no longer runs, where thought
disappears into experience.

Richards presents a more complex case, that of a theorist
of some importance who scuttles his theory as soon as it is
launched. Having elaborated a general theory of value, he
arrives at the conclusion that value is, after all, its own proof,
and theoretical argument is subordinate to and dependent
upon an initial act of pure critical choice made by the 'whole
personality'. And within Richards's theoretical system itself,
what looks like analysis often turns out to be a habit of
numerically subdividing the object into manageable components with no systematic relation established between
them. This is the case most noticeably in the two completely
separate uses of language, but recurs in the four types of
meaning in *Practical Criticism,* and in earlier articles in which
Richards proposes that there are six types of emotion and
'six or eight'[6] uses of the term 'expression' (the magical number
seven having been reserved already, perhaps, for Empson's
ambiguities). A further implicit undermining of theory is
provided when Richards analyses the phenomenon of the
'stock response': the cause of this aberration, Richards insists,
is withdrawal from experience. The recommended cure seems
to require no critical reflection or analysis, but again, further
immersion in experience.

With the Leavises, the distrust for theory, and the cult of
'experience', reach a culmination: they are almost founding
principles of the *Scrutiny* movement, as became apparent in

1937 in the widely discussed dispute between F. R. Leavis
and René Wellek.[7] The early, less adulatory writings on
D. H. Lawrence display this trait particularly well. Lawrence
is held up as a model for his gift of experiencing rather than
thinking: an anti-theoretical judgement which becomes
transformed into an extraordinarily naive critical approach
to literature, taking D. H. Lawrence at his own word in
describing his novels as 'pure passionate experience' [FC, 119,
131]. Apart from ignoring the glaringly obvious discursive
and intellectual elements of Lawrence's novels, this dissolves
all distinction between writing and other activities, since
there is no room for any (necessarily theoretical) distinction
which could break down 'experience' as an all-embracing
category. Like Arnold, Leavis tends to make a virtue of
tautology.

In general, all these critics fill the gap left by their exclusion
of theoretical concerns with a haphazard critical method
loosely combining empiricist and idealist components.[8] Thus,
in Arnold's writings the act of pointing to certain lines of
poetry or to certain social institutions acts as an empirical
anchor for semi-mystical categories like 'charm', 'natural
magic', 'the Celtic spirit', or 'the spirit of the Populace'.
Arnold tends to leap unsystematically from the noumenal
'life' to the phenomenal 'there'; and the same pattern of
short circuits between vitalism and empiricism is repeated
almost exactly by the Leavises. In Richards, the uneasy
mixture of methods is even more visible: the attempt is made
to reconcile an atomistic, semi-behaviourist model of the
human mind as a circle bombarded by particles of stimuli
[PLC, 208], with an antagonistic model of subjective self-
development and critical choice patched together with
Buddhist themes. This forked or double method is an almost
inevitable consequence of the division of the world into sealed
compartments of scientific fact on the one hand and moral
value on the other.

iii. *Literacy and public controversy*

One consequence of the general unease about science and
theory in this critical tradition is a registering of reservations

about the value of mass literacy. For if certain kinds of knowledge are unhealthy or dehumanizing, while 'experience' and cultural equilibrium (both, it would seem, attainable without reading) are preferred to them, then the ironical result can be that defenders of literature are tempted to challenge literature's basic prerequisite, literacy. In the later critics especially it seems that the nineteenth-century debate about whether to allow the mass of the population to read was not finally settled by the Education Act of 1870. Arnold and the authors of the Newbolt Report, to be sure, were certainly committed to mass literacy, as was Richards in his capacity as a promoter of Basic English; but even they tended to view the potential reading-matter of the masses with the eye of the censor: Arnold, in his attack on Bishop Colenso, argued that any books which threatened received ideas should have remained unwritten, untranslated, or unavailable to a mass audience. Richards, similarly, sees the threat of impending chaos as a result of the too rapid and too wide-spread dissemination of new ideas amongst the multitude; the wider the readership, the greater the tendency towards 'levelling-down' in cultural standards. Eliot's remarks upon mute theatre and music halls in his early writings foreshadow his ideal of an illiterate audience for the poet. But the half-formed reservations about mass literacy held by these writers really flourish only in the work of the Leavises.

F. R. Leavis saw Richards's adoption of Basic English as (with some reason) a regression from his former resistance to 'levelling-down'. Q. D. Leavis's *Fiction and the Reading Public* keeps up a constant attack on the supposed advantages of mass literacy, drawing a line between improving 'literature' and mere 'fiction' — which she associates with drug-taking, day-dreaming, and masturbation. This kind of reading simply to fill time and prevent boredom was not necessary to the 'old order' celebrated by Leavis; nor, as she points out, was it possible until the nineteenth century. Long working hours, illiteracy, and the high price of novels were factors which slowed down the rate of cultural decline, and, in Leavis's scale of values, it is difficult not to see these obstacles to a mass readership as something to be thankful for. She is happy to note, for example, that the Elizabethan mob had no choice in its entertainments, but had to take what culture its betters provided. 'The sudden opening of the fiction market to the general public was', she writes, 'a blow to serious reading.' [*FRP*, 133.] F. R. Leavis held the same view, adding in *Culture and Environment* that mass literacy was to blame for the sensationalism and vulgarity of modern newspapers: 'The state of the Press and its power are the unexpected consequences of compulsory education.' [*CE*, 104.] The Leavises' claim that the rural poor before the days of compulsory education had the compensation of a more satisfying existence, 'a way of living that obeyed the natural rhythm' [*FRP*, 169], can be taken only as an implicit approval of mass illiteracy.

As can be seen from many of the remarks on science and literacy by these critics, the ground of their anxiety was that the dissemination of new knowledge would aggravate the spirit of controversy abroad in British life. Arnold's constant complaint was against this factional entrenchment in 'our too political country' [*CPW* iv. 315]. Both his poetics and his politics are an attempt to reach a higher plane, or refuge, of 'disinterestedness' and serenity: his ideal poet is advised to banish from his mind all feelings of contradiction; his myth of Oxford is founded upon the university's supposed distance from fierce intellectual disputes. His is the paradoxical case of a controversialist opposed to controversy, artfully stepping outside the rules of the debate whenever he might be forced to explain himself. The motive for his repeated shiftings, dodges, and vaguenesses is the fear that public controversy would transform itself — as he thought it had done in the Hyde Park disturbances of 1866 — into violence and fanaticism. Under modern conditions, the heat of controversy could no longer be confined to a small circle, as Arnold warned in *Culture and Anarchy*:

We come again here upon Mr. Roebuck's celebrated definition of happiness, upon which I have so often commented: 'I look around me and ask what is the state of England? Is not every man able to say what he likes? I ask you whether the world over, or in past history, there is anything like it? Nothing. I pray that our unrivalled happiness may last.' This is the old story of our system of checks and every Englishman doing as he likes, which we have already seen to have been convenient enough so long as there were only the Barbarians and the Philistines to do what they liked, but to be getting inconvenient, and productive of anarchy, now that the Populace wants to do what it likes too. [*CPW* v. 156-7.]

Under these new conditions, the inflammatory voicing of partisan attitudes must be quelled; hence Arnold's recourse to the power of the state as the only agency able to restrain the clashing of ignorant armies.

In Eliot's writings there is the same impatience with controversy: his ideal society would change unconsciously without theory or polemic. For him, the trouble with English criticism is that, rather than letting the texts speak for themselves, it argues and persuades. The critics appear to him to be like contending orators in a public park, whom he wishes to move along. The virtue of genuine poetry is that in digesting or dissolving thought, it eliminates controversy along with it: if there is a philosophical opinion in a work of literature, it is there, Eliot insists, 'not as matter for argument', but as an incontrovertibly accepted view of the world, almost a physical fact.

With Richards the painless abolition of controversy is a conscious and urgent concern. One of the few critical predecessors singled out by him for admiration, S. T. Coleridge, had written in *Biographia Literaria* of 'the beneficial after-effects of verbal precision in the preclusion of fanaticism, which masters the feelings more especially by indistinct watch-words'.[9] Richards's aim was to elaborate a verbal precision which could disengage the 'indistinct watch-words' or 'pseudo-statements' from their fanatical consequences. This is the intention behind his separation of the emotive and referential functions of language, as he and Ogden announced in their preface to *The Meaning of Meaning*: 'with this understanding it is believed that such controversies as those between Vitalism and Mechanism, Materialism and Idealism, Religion and Science, etc., would lapse'.[10] The same anxiety to dissolve partisan lines of division, and to prevent them from spilling over into fanaticism, is evident in the distinction which Richards draws between the taking of action and the preliminary 'attitude' evoked by poetry.

Although the Leavises have since become a byword for polemical bitterness, their early work, in particular, shows a marked refusal to engage in open controversy. As they often repeat, their arguments are not intended to convince any opponents, but only to rally those already in agreement with them. Against what they regard as divisive and dangerous appeals to the class war, they counterpose the key terms 'consensus' and 'community'. Like Richards, they regard the conventional battle-lines of public strife to be fictions; in particular the struggle between the bourgeois and the Marxist, who are seen as sharing a fundamental community in their reliance upon the machine. The Leavises insisted that *Scrutiny* should identify itself with no particular cause or movement. Yet this refusal only locked their position further into the mould of Arnold's paradox, casting them as controversialists against controversy, rebels whose cause was consensus.

iv. *Whiggery and history*

The difficulty of maintaining such a position against political and ideological partisanship inheres in the fact that these critics found themselves, whether they liked it or not, linked to identifiable political antipathies, if not affiliations. Eliot, most notoriously, spelled these commitments out publicly in the late 1920s, but had already expressed his distaste for 'whiggery' long before. Arnold expounded identifiable political views not just on state education but on Italy and Ireland, which classed him, he felt, as a 'Liberal of the future' — in other words a writer deeply dissatisfied with the Liberalisms of past and present. Hostility to orthodox Liberalism is a strong element of continuity in this critical tradition: the same enemy is identified successively by Arnold as 'Philistinism', by Eliot as 'whiggery', and by the Leavises as 'Benthamism' — a development which culminates in Eliot's assertion that western society is 'worm-eaten with Liberalism'.[11] Arnold's use of ridicule against the short-sighted attachment of the British bourgeoisie to the slogans of free trade and free speech is often light-hearted; but it has a serious aim in attempting to undermine the Liberals' naive faith in automatic progress. Against this faith Arnold and his literary-critical successors revived the political themes of Burke's and de Tocqueville's critiques of democracy. For them, the consequence of liberal democracy is a dilution or levelling-down of cultural standards and an anarchy of competing interests subversive of cultural authority. The arrival of the Populace in the political arena was far more than the 'inconvenience' mentioned by Arnold: for Gosse it presaged a revolt of the mob against their literary masters; for Eliot, Richards, and the Leavises the menace of literary culture's sudden swamping by the values of the semi-literate herd, aided and abetted by Liberal demagogues and the unscrupulous manipulations of the Press.

If this was the relentless onward march of history as projected by the liberal dream of progress, then these literary critics wanted none of it. Arnold's Preface to *Essays in Criticism* greets the inevitably Philistine future in this mock nightmare:

My vivacity is but the last sparkle of flame before we are all in the dark, the last glimpse of colour before we all go into the drab, — the drab of the earnest, prosaic, practical, austerely literal future. Yes, the world will soon be the Philistines'! and then, with every voice, not of thunder, silenced, and the whole earth filled and ennobled every morning by the magnificent roaring of the young lions of the *Daily Telegraph*, we shall all yawn in one another's faces with the dismallest, the most unimpeachable gravity. [*CPW* iii. 287.]

While conceding the immediate future to the Philistines, Arnold identified himself with the spirit of Oxford, with its enduring capacity to keep open its communications with the future by slowly undermining Liberal and Protestant confidence. The really decisive undermining of the 'whig' version of history as a smoothly uninterrupted expansion of British liberty was not to come until the war of 1914–18 — a catastrophe inexplicable in Philistine terms — but when it did come it was enough to open up serious doubts about the simple view of history as gradual betterment, not only in literary criticism, but in criticism none the less. Eliot's anti-historical concept of tradition and the related notions of 'timelessness' were just the most noticeable features of this reappraisal of the idea of history.

To some extent, the groundwork for the post-war assault on history had been laid by Arnold in his lecture 'On the Modern Element in Literature', where he offers a cyclical model of history in which past ages can be more modern than our own — against the 'whig' version's ·linear model of expanding wealth and liberty. For Arnold this meant, alarmingly, that history could go wrong; that it was possible for some historical events, like the French Revolution, to be out of step with the cyclical ebb and flow of spiritual and cultural movement, to be premature. Nevertheless, there is in Arnold still a considerable residue of humanist faith in the future, a conviction that with a little self-control history could right itself. No such faith is held by Eliot, Richards, and the Leavises. Eliot in particular sets up his 'simultaneous order' of poetry in open opposition even to the notion of history as chronological development, and where history is admitted to exist, it is only as degradation or fall, an inversion of the whig interpretation of history with Protestants and Parliamentarians as anti-heroes. The simultaneous or timeless model of literature, at least, became widespread: a noteworthy example being E. M. Forster's *Aspects of the Novel* (1927), in which the author attempts to 'exorcize that demon of chronology' by imagining all novelists 'seated together in a room, a circular room . . . all writing their novels simultaneously'.[12]

Richards and F. R. Leavis both recoiled from history in a very direct sense: both were originally students of History at Cambridge, and both switched courses, Richards to Moral Science, Leavis to English. Richards later recalled that his motives for this change were that he 'couldn't bear history', and that he 'didn't think history ought to have happened'.[13] It has often been noted of Richards's practical criticism that it excludes all historical consideration of literary works. This exclusion arises not from any forgetfulness or purely temporary concentration on textual detail, but from his reaction to history as a profoundly unbearable lapse from a formerly integrated state of mental organization — a reaction quite congruent with Eliot's nostalgia for pre-revolutionary English culture, as in this observation on violence in fiction: 'We may suspect that to-day the demand for violence reflects some poverty, through inhibition, in the everyday emotional life. In Elizabethan times a perhaps not analogous demand could not, however, admit of this explanation.' [PC, 259.] This contrast between an uninhibited, emotionally rich Elizabethan sensibility and an impoverished modern sensibility is the version of historical decline transmitted by Richards from Eliot to the Leavises. For F. R. Leavis, the rehabilitation of seventeenth-century literature in place of that of the nineteenth century was 'the great critical achievement of our time'.[14] It is a radical discontinuity in history (implicit in the title of For Continuity) that the Leavises trace between Shakespeare's national culture, rooted in the soil and appealing to a multi-class audience, and today's empty mechanical civilization. As F. R. Leavis expressed it in Culture and Environment, 'the present phase of human history is, in the strictest sense, abnormal' [CE, 93]. From this, history's cruel anomaly, follows the need for a culture uprooted from the contaminations of modern life and instead rooted, like Lawrence, in the past. Continuity, in this way, could be maintained with the past and communications kept open (the Leavises constantly recall Arnold's phrase) with a remotely possible future in which historical normality is restored.

v. *The problem of order: mental wholes*

The enemies listed above, from science to history, illustrate the common ground shared by Arnold, Eliot, Richards, and the Leavises only negatively. Their positive remedies, values, and catch-phrases share a similar degree of affinity, captured most succinctly in the term Eliot used to convey the continuity of his concerns from 'Tradition and the Individual Talent' to *After Strange Gods*: 'the problem of order'. The term is, deliberately, loose enough to embrace not only literary order but ideological, social, political, and psychic orders, with the boundaries and distinctions between them usually blurred or overlapping. Order or harmony becomes the key term uniting these critics against the disturbing or anarchic implications of science, theory, literacy, controversy, and history; and its weight is constantly reinforced by the attention given to such centrifugal forces propelling the world to chaos. Order is the quality sought in the revival of Coleridge's theory of the imagination as a balance or reconciling of opposites, in Arnold's concept of harmonious human development without premature growths, in the 'equilibrium' of Eliot's account of wit, in Leavis's organic community, and in Richards's insistence (aimed mistakenly against Pater) on the value of organization in experience, rather than of sheer intensity. What is consistently characteristic of this line of critics is less the institutional guarantee of order (which can vary from Arnold's state and Eliot's Church to Richards's League of Nations) than what might be called its subjective correlative: the equation, that is, of social and cultural orders with a certain balance or harmony of the individual mind. If the comprehensive order aspired to by these critics has a core or centre of gravity, it is the individual psyche — almost a First Cause to which literary and social phenomena are referred back for explanation in their work.

Arnold's tendency to collapse together the social and the psychological has been noted at some length in Chapter 2; particularly his use of the term 'best self' to describe the state, and his identification of the violent 'spirit of the Populace' with a psychological streak of brutality present in all of us. His entire conception of 'Culture' rests upon similar analogies and conflations, achieved via the central idea of the harmonious development of human capacities (as in an individual's growth and education) projected into other spheres. In his work, it is the character-forming properties of the literary 'grand style' and of school education which are picked out for emphasis. Likewise Arnold's writings on religion stress the same process of internalization and personal identification as his writings on the state. Referring to Isaiah and the Psalmist in *Literature and Dogma* as precursors of Christ, Arnold asserts: 'This is a personal religion; religion consisting in the inward feeling and disposition of the individual himself, rather than in the performance of outward acts towards religion or society. It is the essence of Christianity . . .' [*CPW*, vi. 217]. As with Arnold's concept of 'culture', his religion and his politics can be said to be essentially an 'inward condition' — an inwardness taken by Pater as the cue to contract culture to self-culture.

One prominent characteristic of the early Professors of English Literature, noted in Chapter 3, was their reduction of literature to a matter of individual character or mentality, in even more pronounced fashion than Arnold. For Charles Kingsley, literature offered — particularly to women — a training in personal sympathy with authors and fictional characters, while Caroline Spurgeon regarded poetry as above all a heightened state of mind rather than a certain kind of literary work with its own procedures; and Raleigh, most ostentatiously, interested himself only in the 'live men' of literature behind the litter of paper documents. The arrival of T. S. Eliot's 'impersonal' theory certainly challenged this kind of critical approach, re-establishing an idea of literature as literature rather than as psychology or biography; but it did not mark a complete break from 'psychologistic' approaches to literature and to history. Eliot's theory of the dissociation of sensibility, in particular, offered a version of English literature and English social and political history in terms of mental integration, swallowing that literature and that history into 'the English mind'. In this account of history, the fusion of thought and feeling in a balanced 'sensibility' is set up as a model of mental order against which literary works and historical events are to be judged.

In Richards's work, the reduction of all questions to the question of mental organization is carried to its furthest limit. He goes as far as to claim, for example, that 'Far more life is wasted through muddled mental organisation than through lack of opportunity. Conflicts between different impulses are the greatest evils which afflict mankind.' [SAP, 33.] The same merging of the microcosmic level of individual psychology with the macrocosm of world politics (and the same overloading of the term 'impulse') appears in his plea for 'a League of Nations for the moral ordering of the impulses' [SAP, 35]. All human problems become for Richards problems of mental health, with art as the cure: taking 'the natural isolation and severance of minds' [PLC, 137] as his starting-point as Pater had done, he argues that a poem or work of art *is* the state of mind which it produces or which produces it. Literary criticism then becomes a question of attaining the

right state of mind to judge — other minds, according to
their degree of immaturity, inhibition, or perversion. The
paradoxical result of the 'practical criticism' approach in its
exclusion of biographical data is to place more, not less
importance on certain implicit personal characteristics of
the author or even of the reader. For Richards, it is 'sincerity'
that 'is the quality we most insistently require in poetry. It is
also the quality we most need as critics.' [*PC*, 282-3.]

The Leavises continue this search for qualities of personal
integrity in or through literature, usually employing the term
'maturity' rather than 'sincerity'. Characteristic of this
inheritance from Richards is F. R. Leavis's assertion that
D. H. Lawrence represents 'in his living person' the human
tradition and human normality, and Q. D. Leavis's relegation
of most fiction to the status of 'cases' for (presumably
abnormal) psychology rather than for criticism. Novels are,
for the Leavises, more or less simple reflections of the
mentality, mature or immature, of their authors; advertisements direct projections of the mentality of the herd. In the
same way, a society can be understood almost solely by
looking at its arts: the unsegregated audience of Elizabethan
drama comes to stand for a supposed unity of Elizabethan
'life' as a whole. By their endorsement of Ezra Pound's
argument for the importance of literature in the State, the
Leavises even maintain that entire societies stand or fall by
the cleanliness of their tools of thought.

George Watson has pointed to 'the almost total absence of
psychological criticism'[15] in the early twentieth century as
an astonishing paradox, given the growth of this new science
at that time, and I. A. Richards's apparent familiarity with
it. To unravel the contradiction here, it is helpful first to
distinguish (as Watson fails to do) psychology in a loose and
general sense from psychoanalysis. It is certainly true that the
Leavises were hostile to the latter; F. R. Leavis complaining,
for example, that 'psychoanalysis has been, for those
interested in literary criticism, merely a nuisance'.[16] Yet
their critical concern for the health of the author's mentality
can still be characterized as psychological in the loose sense,
with the stress on the psyche rather than on the logic. With
Richards, the distinction cannot be so clearly drawn, since

he defined his position within psychological theory as that of a 'centrist' between the 'extreme wings' of behaviourism and Freudian psychoanalysis, preferring to follow those 'cautious, traditional, academic, semi-philosophical psychologists' [PC, 322] who draw upon both wings. As an attempt at conciliation this was as certainly doomed to failure as the League of Nations itself, and as an account of Richards's position, it is difficult to square with the evidence of his major writings, all of which show a much stronger behaviourist than Freudian influence; but the insistence upon caution is in itself a useful indicator of his disquiet. In the concluding paragraph of *Science and Poetry,* immediately before his remarks on the impending mental chaos from which poetry can save us, Richards has this to say about the new science:

In many quarters there is a tendency to suppose that the series of attacks upon received ideas which began, shall we say, with Galileo and rose to a climax with Darwinism, has over-reached itself with Einstein and Eddington, and that the battle is now due to die down. This view seems to be too optimistic. The most dangerous of the sciences is only now beginning to come into action. I am thinking less of Psycho-analysis or of Behaviourism than of the whole subject which includes them. [*SAP,* 82.]

The danger which this new science poses to received ideas lies in its challenge to the supposed unity and wholeness of the mind; hence the 'mental chaos' which Richards expects to be brought about, and hence his own ambiguous attitude to psychology. He and the Leavises are 'psychological' critics to the extent that they attempt to sustain and exemplify a model of psychic order as a means of defining and measuring the attainments of literature. To the extent that modern schools of psychology undermine that model (as Freudian psychoanalysis, in particular, does), they regard the subject as a nuisance and a danger.

In this context it is instructive to compare Freud's epoch-making *Three Essays on the Theory of Sexuality* (1905) with the assumptions of English criticism. The explicit premise of Freud's most important work was a dissolution of the accepted boundaries of normality and 'perversion' — boundaries which to these critics seem to have been indispensable principles. There is a remarkable continuity here from Arnold's 'eccentricity' and 'ordinary self', through Eliot's 'impure desires' and 'heresies', and Richards's 'perversions' and 'Narcissism', to the revulsion of the Leavises from adolescence, masturbation (another black mark for Leopold Bloom), and 'immaturity'. In their scale of literary judgements this encourages the Leavises to elevate the heterosexually 'fecund' D. H. Lawrence as a totem against the reviled homosexual W. H. Auden. An inflation of the super-ego is perhaps inescapable in critics in so far as they act as 'judges'; but this line of critics is not only judicial in tone but positively inquisitorial, indulging in a kind of perversion-hunting quite at odds with Freudian thinking.

A full explanation for the absence of a Freudian criticism in Britain in the 1920s and 1930s would require a more extensive investigation into Freud's reception in British intellectual life as a whole. But from the limited standpoint of literary history a preliminary proposition may be hazarded: that the biographical traditions of nineteenth-century criticism, combined with Eliot's ideal of the undissociated sensibility, had so 'inoculated' English criticism with psychological assumptions that Freud's work could not be assimilated lest it undermine some of that criticism's most essential categories. A simple model of normality and mental consistency was too ingrained an assumption to be given up. Upon it depended, for instance, many of the premises of the literary critique of 'mass culture', in particular the idea that one mind cannot take in more than one kind of cultural training, propounded here by Q. D. Leavis:

The training of the reader who spends his leisure in cinemas, looking through magazines and newspapers, listening to jazz music, does not merely fail to help him, it prevents him from normal development ... partly by providing him with a set of habits inimical to mental effort. Even in small matters it gets in his way: for example, the preconceptions acquired from the magazine story and the circulating library novel are opposed to any possibility of grasping a serious novelist's intention. [*FRP*, 180.]

Such a conception of unbroken and exclusive identity between the mind and its surrounding cultural influences is essential to these critics' contention that 'mass culture' is not only inferior but actually damaging. More important still is the maintenance of the doctrine of psychic wholeness in and through literature as an analogue for a projected harmony and order in society.

vi. *The problem of order: the social organism*

The close parallels pursued by these critics between a mental and a social order and equilibrium are not, of course, particularly original: they have a wealth of precedents in European thought going back, as Richards later pointed out, to 'the founding metaphor . . . of Western philosophy',[17] the mind/ community analogy of Plato's *Republic*. The analogy is enforced by these critics, moreover, with a further well-worn metaphor, that of the body politic. That is to say, they supplement the idealist projection of mental order with a more positivist theme of necessary or *natural* order and consistency. These twin models find their expression in the couplet of explanatory terms which Sainte-Beuve had proposed as the keys to any author's literary work: character and race. The importance of character for these critics has been commented upon above. In tracing the connection which they imply between mental and social orders, the intermediary concepts of race and national character will need more examination than most apologists for this critical tradition have been prepared to undertake.

Arnold's writings are permeated with assumptions about the common racial basis of literary and social phenomena, most obviously indicated by his vocabulary of Hebraism, Hellenism, and Philistinism. The tendency at work in many of his arguments is an urge to 'naturalize' human activities and institutions, reducing them to expressions of an irresistible natural law. Thus, his explanation of ethics in his religious writings is founded upon a tendency in all creatures to fulfil the laws of their being. Arnold's objections to sudden and premature social changes are similarly given a 'natural' pretext in *Culture and Anarchy*: 'Everything teaches how gradually nature would have all profound changes brought about; and we can even see, too, where the absolute abrupt stoppage of

feudal habits has worked harm.' [*CPW* v. 205.] He even
proposes in 'The Modern Element in Literature' what might
anachronistically (since it was written before 1859) be called
a literary Darwinism, whereby lost classical works are deemed
to have been rejected by 'the instinct for self-preservation
in humanity' [*CPW* i. 29] which abandons those literary
works, such as Menander's, whose frivolity fails to foster
human life. He devoted an entire work, *The Study of Celtic
Literature*, to tracing the assumed racial roots of British
poetry back to their Celtic, Norman, and Germanic sources.
In this work, Arnold goes still further and criticizes English
and German hymns for their misguided attempt to imitate
Semitic cultural forms: 'we are none the better for wanting
the perception to discern a natural law, which is, after all,
like every natural law, irresistible; we are none the better for
trying to make ourselves Semitic, when Nature has made us
Indo-European, and to shift the basis of our poetry.' [*CPW*
iii. 369.]

Similar opinions about literature's (and criticism's) necessary conformity to racial character recurred in the bitter
battles over 'Teutonism' in the early English studies movement, culminating in the wartime resistance to 'alien tyrannies'
and in such assertions as Quiller-Couch's claim that Germans
were constitutionally incapable of reading English poetry. One
of the side-effects of this trend was to encourage the growth
in the early years of the twentieth century of a nostalgic
Elizabethanism, both in the openly jingoistic verse of Henry
Newbolt and in the more restrained critical revaluations
promoted by Eliot.

Even Richards, usually careful to distance himself from
the heated chauvinism of the war, can be seen to absorb
certain related assumptions along with his positivist inheritance,
as in this qualification to his general theory of value: 'Within
racial boundaries, and perhaps within the limits of certain
very general types, many impulses are common to all men.'
[*PLC*, 148.] Once the racially specific foundation for
'impulses' is assumed, it works its way into Richards's theory
of cultural decline as a caution against excessive mixing of
cultures. Thus in *Practical Criticism*, Richards argues that our
attitudes as speakers, writers, readers, or listeners are in rapid

decline, as a result of 'the mixtures of culture that the printed word has caused' [*PC,* 339]. In such times of dangerous confusion, Richards envisages 'the mind of the future' which can adapt or adjust itself to a variety of impulses without confusion, as a factor of no less than evolutionary significance, the poet or 'superior man' leading a biological mutation of the species.

The same kind of pseudo-Darwinism is employed by the Leavises, notably in *Culture and Environment,* where teachers of English are hailed as the representatives of humanity's instinct for self-preservation and health — a straightforwardly Arnoldian formula. Related ethnological implications are involved, too, in the category of the 'herd' which the Leavises inherit from William Trotter's biological version of sociology. In both its 'organic' and 'herd' forms, the Leavises' concept of human community is 'naturalized' by an insistence upon physical and biological foundations: the strength of Shakespeare and Bunyan is seen to derive from a national culture rooted in the soil, and capable of stirring the blood with 'the accumulated religious associations of a race' [*FRP,* 90]. The decline from the wholeness of old English rural communities and their language is seen as a result of a dilution of Englishness either by Miltonic Latinism or by Hollywood Americanism. The Leavises' idealized community of the old English countryside is nationally cohesive, self-consistent, and organic.

The importance of the twin concepts of race and character in this tradition of literary criticism lies in their potential to function as the bases for a cultural 'sociology' or implicit social model at odds with the conventional conceptions of class warriors. This potential and its consequences for social analysis are most abundantly displayed in the policy statement of the short-lived *Calendar of Modern Letters* (1925-7), an important forerunner of *Scrutiny* in its critical approach and vocabulary. The statement announced:

The reader we have in mind, the ideal reader, is not one with whom we share any particular set of admirations and beliefs. The age of idols is past, for an idol implies a herd . . . and for the modern mind the age of herds is past. . . .

To-day there is only the race, the biological and economic environment; and the individual. Between these extremes there is no class, craft, art, sex, sect or other sub-division which, it seems to us, can claim privilege of the rest. It is with the mind of the individual . . . that literature communicates. Perhaps there are not so many individuals as there are men and women with names and addresses. Perhaps the streams of people in the street are no more dissimilar than autumn leaves, manure for next summer's generations. . . . It is through him [the artist] that we can perfect our individuality, our own shape . . .[18]

This states with uncommon clarity the basic tenets of the implicit social analysis maintained by Arnold and his followers, in particular the irrelevance of intermediary social categories like class, sex, creed, and occupation, between the all-important categories of race and individual character. Society is seen as an aggregation of unknown faces, creatures of the street, herds without character, against whose anonymity only art can forge an individuality from the available racial material.

Arnold's social commentary constructs just such a pattern: undermining the usual notions of class structure by treating the three classes as tribes or spirits, he attributes equal blame to each for its narrow selfishness, offering as an alternative progressive social agency the enlightened 'aliens' or 'remnant' who remove themselves from the passing enthusiasms of the crowd or sect. These individuals are remarkable for their 'humanity', and for their personal identification with the authority of the state as an extension of their 'best self'. Arnold's purpose was to eliminate conceptually those intermediary social institutions and loyalties which obstructed the full communion of individual and state. Positing the existence of 'a *whole society* that has resolved no longer to live by bread alone',[19] Arnold can proceed to erase divisive economic and political issues and in their place to reconstruct a model of society around the unifying centre of cultural habits and relations. Similar attempts to dissolve social divisions into cultural unity recur in the early English studies movement — for example, George Sampson's concept of an 'immaterial' communism whereby true equality is achieved at the opera-house, or Henry Newbolt's appeal to the English Association to help people 'forget the existence of classes' in favour of national unity.

Eliot and Richards share a relatively simple model of society divided between the chaotically self-willed 'large crawling mass' and the cultured vanguard of the species which comprises at one time 'half a dozen men' and at others 'about three thousand people in London and elsewhere' [SW, 121]. For both critics, the spread of the cinema threatens to disrupt the emotional life of the masses, reducing them to a state of helpless suggestibility. The Leavises expand a similar model in much more detail, using the same concepts of levelling-down and media hypnotism, but extending the threat to include 'the machine' as a whole, organized in a 'power economy' hostile to family life. The absolute malleability of the mass or herd is a point they repeatedly insist on, emphasizing as it does the power of a small circle of press and advertising 'fixers' and the corresponding obligations of what could be called a counter-élite of critics and teachers. With the concept of class rendered redundant outside the strictly economic sphere, the Leavises' social model revolves around the opposition between society at large and 'society' in the eighteenth-century sense; between an unconscious mass and its conscious embodiment or guardian. As F. R. Leavis observed in 1963: 'It was essential to the conception of Scrutiny to demonstrate that . . . a public could be rallied, a key community of the élite (one therefore disproportionately influential) formed and held.'[20] Yet this élite could not be expected to emerge from its traditional source in the ruling class, since this class too had become corrupted by the very standardizing tendencies which it had allowed to take root. The public school conformist had appeared. The élite, then, had to be drawn, like Arnold's 'aliens', from outside the major forces of class-bound British society; from those typically 'independent' social groups, the small farmers and shopkeepers whose traditions the Leavises admired, and from their sons and daughters in the teaching profession.

The social model constructed by critics in this tradition is usually 'implicit', but it is necessary to the criticism, and to the priorities and particular urgencies propounded in the criticism, even where the sociological assumptions are not spelt out in detail. In general, as this school of literary criticism claims a larger importance in the world, it is forced to incorporate within its discourse larger and larger assumptions about the construction and working of the society in which it is active, and to distort the elements of its social analysis to allow criticism itself a decisive position. The result in terms of sociological theory is a certain 'culturalism', that is to say an undue assumption that a society is constituted and changed chiefly by its cultural features, forms, and institutions, that this or that use of language or 'sensibility' can save or condemn a civilization regardless of the 'machinery' through which it must operate. The reiterated term 'machinery' itself becomes the derogatory indicator of an entire secondary underworld of political institutions and economic life which is of importance only as it expresses or fails to express the inner, the essential character of a culture.

vii. *Substitution and ideology*

It is the substitution of literary-critical discourse for 'culture' in general and for a whole range of other discourses from the philosophical to the political and sociological that most characterizes the general trend of this period in English criticism. The 'expansionism' whereby criticism conceptually abolished and took over other spheres was announced boldly in Arnold's 'The Study of Poetry', where the prospect of poetry replacing religion was advanced. Yet more was at stake in this critical project than merely poetry and religion, important though they were; what is developed in this tradition is an attempted substitution of literary criticism for the kinds of competing activities discussed earlier in this chapter. In Arnold's work this attempt is made, for example, when he proposes that criticism is the guardian of a 'general culture' in whose interests it judges all books over and above their specialist importance. Arnold's projected substitution is not simply the replacement of religion by poetry, but an accompanying operation performed between their respective 'dismal sciences', criticism and theology.

Criticism, according to Arnold's redefinition of it as the attempt to know the best that has been thought and written, becomes a monitor of 'general culture' as philosophy or theology had been before. It is even smuggled into poetry as

Arnold conceived it: his much-debated assertion that poetry is a 'criticism of life' has the effect (whatever else it may mean) both of making poetry aspire to the condition of criticism, and of making criticism's object 'life' itself at only one remove. Richards in particular was quick to notice, and to repeat this effect. His reaffirmation of Arnold's position appears, significantly, as part of his most vehement defence of criticism, not, as one might expect, in a simple defence of poetry:

But it is not true that criticism is a luxury trade. The rear-guard of society cannot be extricated until the vanguard has gone further. Good-will and intelligence are still too little available. The critic, we have said, is as much concerned with the health of the mind as any doctor with the health of the body. To set up as a critic is to set up as a judge of values. What are the other qualifications required we shall see later. For the arts are inevitably and quite apart from any intentions of the artist an appraisal of existence. Matthew Arnold when he said that poetry is a criticism of life was saying something so obvious that it is constantly overlooked. The artist is concerned with the record and perpetuation of the experiences which seem to him most worth having. [*PLC*, 46.]

Richards appears to be transferring his attention here from critics as judges of value to artists as judges of value, yet the true focus of his interest is betrayed by the implausibility of the last, Shelleyan, sentence in this passage. (A poem such as *Paradise Lost,* for example, is most certainly an appraisal of existence, yet it records precisely the experience least worth having.) It would be much more accurate to say that 'the critic is concerned with the record and perpetuation of the books which seem to him or her most worth having', since it is really criticism and the extent of the critic's powers which concern Richards here. The definition of the task of the arts is drafted so as to allow the easiest short cut between criticism and life, putting all 'values' and 'experience' in general at the disposal of the critic. The extent of the critic's domain is illustrated elsewhere in *Principles of Literary Criticism* when Richards refers to 'such topics as truth in art, the intellect-*versus*-emotion *imbroglio,* the scope of science, the nature of religion and many others with which criticism must deal' [*PLC*, 96].

Richards's literary theories result in an enhancement of the value less of poetry than of criticism itself, as a superior exercise of human faculties, a crossroads in the development of personality. This claim was equivalent to the educational ambitions of the Newbolt Report, with its vision of English as the 'keystone' to the arch of national education; and of course to the Leavises' conception of English Literature as a 'central' subject at university level. For the Leavises, indeed, the university School of English could become the 'centre' of civilization itself, a kind of paradigm of the recreation of life, as F. R. Leavis explained in his 'Retrospect' to *Scrutiny*:

Our special business was literary criticism but we saw nothing arbitrary in taking the creative process of criticism — that interplay of personal judgements in which values are established and a world created that is neither public in the sense congenial to science nor merely private — as representative and type of the process in which the human world is created and renewed and kept living . . .[21]

What he meant by this is perhaps more easily understood in Leavis's assertion in *Culture and Environment* that literary study is a substitute for a whole lost world, that of the 'organic community'.

That this line of English critics had 'expansionist' ambitions for their craft is noticeable enough; what is less easy to define with any precision is the nature of the terrain they were setting out to colonize. Their detractors are usually content to describe the area as the 'moral', or to speak in terms of substitute religions. But there is, even in the short space of Arnold's 'The Study of Poetry', more than this: Arnold speaks of replacing creeds, dogmas, philosophies, received traditions, consolations, and interpretations of the world, and of the incompleteness of science. F. R. Leavis pointed to a related constellation of elements in *Culture and Environment* when discussing language and tradition: 'The decisive use of words to-day is, as we have seen, in association with advertising, journalism, best-sellers, motor-cars and the cinema.' [*CE*, 82.] With the possible exception of motor cars, these are the areas where 'values' are transmitted and renewed or degraded. For Leavis they are enemy territory, to be neutralized if literature and criticism are to advance their claim to be (as

Pound had said they must be) themselves the 'decisive use of words'.

Yet another list is offered by Richards in *Practical Criticism*: 'The methods I have here tried to apply to critical questions, have to be applied to questions of morals, political theory, logic, economics, metaphysics, religion and psychology . . .' [*PC*, 341]. Some rationale for this rather daunting programme of enquiry appears in the introductory chapter of the same work, where Richards maps out of three kinds of discussion: on the one hand science and mathematics, which can be discussed in strictly verifiable terms, on the other hand practical organizational matters which can safely be left to established convention and rules of thumb (Richards includes commerce, law, and police work here), and in the middle the area of unverifiable opinion and feeling associated with terms like morals, ethics, justice, liberty, faith, love, and religion — a whole world of assumptions, guesses, and prejudices.

As a subject-matter for discussion, poetry is a central and typical denizen of this world. It is so both by its own nature and by the type of discussion with which it is traditionally associated. It serves, therefore, as an eminently suitable *bait* for anyone who wishes to trap the current opinions and responses in this middle field for the purpose of examining and comparing them, and with a view to advancing our knowledge of what may be called the natural history of human opinions and feelings.

In part then this book is the record of a piece of field-work in comparative ideology. [*PC*, 6.]

The concept of 'ideology' may be as notoriously inexact as what it denotes,[22] but it is still useful to adopt Richards's term, if only to provide a common label covering a particular area of discourse upon which these critics focus without agreeing on its name: Arnold's *Aberglaube*, Eliot's 'tradition', Richards's 'pseudo-statements' or 'attitudes', the Leavises' 'decisive use of words' or 'consensus of values'. The area is clearly wider than religion, encompassing values, habits, opinions, and assumptions of all kinds; within it, Richards argues, poetry and the discussion of poetry are 'central', and it is this position of dominance within the ideological world as a whole that this line of critics seeks to enforce.

If 'ideology' be adopted as a single term to cover that

realm of values in which criticism and poetry are seen to operate, then many of the common concerns of Arnold and his successors come into clearer focus. What seems to pre-occupy these critics is the system of largely unconscious habits and assumptions necessary to social coherence, which poetry and criticism rely upon and renew. It is the ideological suppleness derived from these sources which equips poetry to replace the old 'feudal habits of subordination' (for Arnold) or 'organic' ties and loyalties (for the Leavises) whose dis-appearance threatens a rapid hardening of social relations into a dangerous brittleness. The ability of poetry and of criticism to bypass the rigid positions of philosophies and creeds is, given such a danger, a jealously guarded asset. Hence, for example, Eliot's insistence that works of art must deal with already accepted ideas or transmute philosophy into an incontrovertible 'vision'. Eliot's preference for Elizabethan and Jacobean authors, and for Dante, derives partly from the 'assumption of permanence' [SE, 203] which he finds in their work but not in the literature of more self-conscious and unsettled periods.

Even after recoiling from Arnold's position and embracing dogma and orthodoxy, Eliot still insisted upon the importance of literature for the unconscious, undoctrinal maintenance of cultural order: 'What I want', he wrote in 'Religion and Literature' (1935), 'is a literature which should be *uncon-sciously,* rather than deliberately and defiantly, Christian'. [SE, 392.] It is at this same level of unconscious assumptions that the Leavises sought to reconstitute the consensus that was available to past literary critics. They both point out that Arnold's key terms, 'right reason', 'the best self', and others, assume a level of agreement among a large audience. In his essay 'What's Wrong with Criticism?' F. R. Leavis asserted that in Samuel Johnson's time 'it never occurred to anyone to question that there were, in all things, standards above the level of the ordinary man' [FC, 89]. The enormous importance of literary criticism for modern civilization was that it presupposed and developed a 'consensus' which had been lost in all other social spheres and relationships. Criticism characteristically defines its own audience in an act of corroboration between minds already in substantial agreement.

In the Introduction to his anthology of criticism from the *Calendar of Modern Letters* F. R. Leavis defended the lack of definition in the *Calendar*'s standards: 'These "standards of criticism" are assumed; nothing more is said about them. Nothing more needed to be said about them; for if we can appreciate — which is not necessarily to agree with — the reviewing of *The Calendar,* we know what they are, and if we cannot, then no amount of explaining or arguing will make much difference.'[23] From this closed circle of assumptions, the Leavises aimed to renew 'that basis of things taken for granted which is necessary to a healthy culture'.[24] Literary criticism was 'central' precisely because it knew how to take things for granted (it could, as it were, take its own taking-for-granted for granted), while other areas of ideology had fallen victim to creeds and abstract definitions. The 'implicit' consensus involved in criticism could offer a model of unspoken community to pit against the prevailing disintegration of culture.

The general project of Arnold and his followers can be described as an attempt to replace the current dogmatic and explicit forms of ideological expression with the implicit and intuitive properties of literary 'sensibility'. Such a substitution was seen to have two important advantages: flexibility and affectivity. As Arnold insisted, literary discussion allows one to shift one's ground unnoticed, avoiding the deadlock of controversy and the revolutionary terror of that logical 'guillotine' which overshadows other kinds of discourse. For him, criticism's most important quality was a certain poise which it required nimble footwork to maintain. Richards was making the same point when he warned readers of *Practical Criticism* to be 'prepared to find little argumentation in these pages, but much analysis, much rather strenuous exercise in changing our ground and a good deal of rather intricate navigation. Navigation, in fact — the art of knowing where we are wherever, as mental travellers, we may go — is the main subject of the book . . . criticism itself is very largely, though not wholly, an exercise in navigation.' [*PC,* 10–11.] From these flexible properties of literary criticism, Richards projected the 'mind of the future' which will be able to ride out the coming ideological turmoil brought on by democracy and the collapse of religious sanctions. Trained in sensitivity to literature, such a mind would be accustomed to navigate as in an aeroplane rather than cling to the 'rock' of inflexible ideological retrenchment. The practical importance of such flexibility was to be its ability to construct an 'innocent language' capable of side-stepping and possibility eroding the rigid lines of social conflict. The study of elevated literary models, it was claimed, could soften the stridency of self-interest, leaving some leeway for humility and self-control. In literature as in religious ideology, Arnold sought to replace repressive and restrictive rules, which only invited rebellion, with ennobling examples and models which invited conformity through self-restraint, conscience, and the subordination of a 'lower' to a 'higher' self. Most graphically, H. G. Robinson offered the image of the clown in the boudoir, stricken by shame into an effort of self-cultivation. The study of literature could have a 'homeopathic' effect by offering its students the passion and grandeur of great literary works so as to induce a sense of modesty and inadequacy — a sense further elaborated by Richards's notions of 'immaturity' in most readings of poetry. The right kind of deference could more readily be commanded by affective literary works than by an external, alien authority: Arnold used the example of Shakespeare and the Church of England's Thirty-nine Articles — Shakespeare, he insisted, was more 'stable'. The problem for critics coming after Arnold was that even this stability seemed under threat from stubborn antagonists: the Newbolt Report noted that the working class was contemptuous of literature, and Edmund Gosse expressed the fear (later quoted by the Leavises) that there would be 'a revolt of the mob against our literary masters'. Gosse was particularly aware of the fragility of the popular deference upon which literature's status rested:

[Lovers of poetry] would produce no general effect at all if they were not surrounded by a very much larger number of persons who, without taste for poetry themselves, are yet traditionally impressed with its value, and treat it with conventional respect. . . . All this beautiful pinnacled structure of the glory of verse, this splendid position of poetry at the summit of the civil ornaments of the Empire, is built of carven ice, and needs nothing but that the hot popular breath be turned upon it to sink into so much water. . . . To speak rudely, it is kept there by an effort of bluff on the part of a small influential class.[25]

If it was a bluff, then the soliciting of conventional respect for literature was still a bluff of great importance for those who had staked the very coherence of society upon it, as the last refuge of 'higher standards' in a period of declining religious certainties. Upon it also depended the claims made for the social effectiveness of literary studies; if these were to carry any credence, then, as Richards and the Leavises perceived, the maintenance of conventional respect had positively to be renewed and reinforced.

The attempted substitution of criticism and poetry for religion in this period of English letters can be, and usually has been, approached as a relatively simple adoption by critics of religiose habits and gestures. Such approaches have commonly been satisfied with the derisory use of the vocabulary of 'sects', 'crusades', 'high priests', and 'disciples', usually against the *Scrutiny* movement. Whatever the temptations — and the Newbolt Report at least invited such treatment — this approach cannot constitute more than a merely frivolous consideration of the complex relations between criticism and the decline of religion;[26] the omnipresence of Christian terms in the vocabulary of European culture makes this kind of exercise too easy to mean anything. The true extent of the substitution is greater than can be accounted for by the simple description of criticism as a 'substitute religion'. It is less religion as religion than religion as occupant of a privileged 'pinnacle' in relation to other kinds of ideology that Arnold and his followers tried to replace with literary discourse, creating a substitute moral philosophy and a substitute social analysis as much as a substitute religion. Poetry was to become a kind of lynchpin for a whole range of other social habits, moral values, and assumptions, confirming them and reflecting them back in harmonized, self-consistent, and emotionally appealing form. If it could fill the gap vacated by religion, literature could offer its own principles of internal consistency, completeness, and regularity of form as a shaping and governing principle for all the conscious and unconscious affairs of society. The order of the one and the order of the other would fall into an

'organic' congruity, a harmonious, rounded, and self-complete
development of civilization under the guardianship of literary
criticism.

Such, at its most ambitious, was the stated or unstated
tendency of the constant analogies between social and
literary orders in this period of English criticism. In usurping
the place of religion as guardian of beliefs and values, though,
criticism could not expect a smooth succession. In particular,
it could no more escape its own schisms and heresies than
could the Church before it; and 'high church' and 'low church'
currents appeared within literary criticism itself. Arnold's
feeling for the nobility of the Catholic Church and his view
that great works are achieved only by Establishments, was
developed by Pater in the direction of ritualism and by Eliot
in the direction of dogmatic 'orthodoxy'. Against them were
ranged the more populist and 'low church' Anglophile
tendencies of the Newbolt Report, of the Leavises, and
of the Quaker and pacifist elements in Richards's work.
To Eliot, the attempt to substitute poetry for religion
appeared increasingly to be a dangerous failure, merely
reproducing the Protestant and Liberal individualism of
the 'inner voice' (as for example in F. R. Leavis's championship of D. H. Lawrence) which Arnold had set out to oppose.
Ultimately a political divergence made itself felt between
the high-church Eliot and the nonconformist Leavises which
stretched back to the century whose literature they valued
in common: for the Leavises the Revolution of the seventeenth century was regrettable in its failure to establish an
independent peasantry which could resist or outlast the
growth of industry,[27] while for Eliot it was regrettable in that
it ever took place at all. Criticism failed to erase the marks of
such divisions in English intellectual life, and proved incapable
of offering that harmonious and conciliatory example which
had been claimed for it. On the contrary, from the 1930s
and beyond, controversy and partisanship in English
literary criticism seemed to flourish almost in direct proportion to the extent of its claims to non-partisanship and
disinterestedness.

viii. *The practicality of criticism*

The very real divergences between the positions of Arnold, Eliot, Richards, the Leavises, and the authors of the Newbolt Report need not be artificially played down in order to enforce a continuity between them. That common ground is substantial enough to allow many partially conflicting approaches without casting into doubt their shared commitment to 'practical criticism'; to the importance, that is, of literary criticism as a discipline of considerable practical importance for civilization at a turning-point. The term 'practical criticism' itself has been flexible enough to accommodate variety within unity, and the history of its redefinition at the hands of these critics can be instructive to students of literary criticism, reminding us to consider both tendencies together.

Coleridge's use of the term in *Biographia Literaria* had implied a definite and logical expository procedure. The fifteenth chapter begins: 'In the application of these principles to purposes of practical criticism as employed in the appraisal of works more or less imperfect, I have endeavoured to discover what the qualities in a poet are, which may be deemed promises and specific symptoms of poetic power . . .'.[28] The principles to which Coleridge refers are those expounded laboriously in the previous chapters of *Biographia Literaria,* and 'practical criticism' refers here to the application of a previously elaborated theory to specific examples. For the notoriously 'theoretical' Coleridge, practical criticism was a procedure subordinate to critical theory and to philosophy, not antithetical to them. At the very heart of Arnold's major innovations in English criticism is his reversal of Coleridge's position on this point. For him, literary criticism becomes an *alternative* to philosophy, logic, and theory. Yet, as was noted in Chapter 2, Arnold used the term 'practical criticism' to designate that undesirable form of discourse which ties itself to particular party or class interests in public polemics. It was in avoiding implication in the sordid business of 'practical' affairs that Arnold felt his 'disinterested' criticism would be ultimately most beneficially influential and (in that sense) practical.

Eliot uses the term in a different sense again, when he requires the critic to work strictly for the improvement of artistic practice, his ideal being the practitioner-critic. He refers, for example, to Ben Jonson thus: 'It is rather in his practical criticism — I mean here not so much his criticism of individual writers, but his advice to the practitioner — that Jonson made progress.'[29] This is already an infringement, on behalf of minority interests, of Arnold's aloofness from the world of practice, although Eliot otherwise avoids implying a wider utility for criticism. The fact that Eliot had in this instance to specify his use of the phrase was due to the currency of another, now established sense of 'practical criticism': the sense given to it by Richards's book of the same name a few years earlier. Here 'practical' is used partly in the sense used, for example, by chemistry students: a carefully designed pedagogical exercise upon particular specimens. The exercise is recognizably distinct from the study of theory — so distinct in fact that Coleridge's sense of the practical application of an already finished theory no longer applies even remotely, and the chemical analogy breaks down: the students enter the laboratory, as it were, blindfolded, and the labels are removed from the bottles. Shadowing this sense in Richards's work is a further sense of the term which makes criticism 'practical' by virtue of its effect in the world at large — its capacity to save a disintegrating culture and to heal mental imbalance. Practicality here is opposed not so much to theory as to the notion of criticism as a luxury trade. It is this sense, though never fully detached from the theory–practice opposition, that F. R. Leavis sought to stress, against what appeared to him to be suspiciously scientistic connotations of Richards's usage. The clearest formulation of this sense appears in his 'Retrospect' to *Scrutiny*:

I mean, we had no tendency to confine ourselves to questions of method or theory, and the 'practicality' of the 'practical criticism' we were indeed (taking 'practical' as the antithesis of 'theoretical') concerned to promote was not just a matter of analytic technique and brilliant exercises. What governed our thinking and engaged our sense of urgency was the inclusive, the underlying and overriding preoccupation: the preoccupation with the critical function as it was performed, or not performed, for our civilisation, our time, and us.[30]

This kind of practicality is in its directness and urgency far from the 'disinterested' ideal promoted by Arnold, yet it is none the less fully in Arnold's tradition, concentrating sharply a concern for criticism's social importance which had long remained latent or 'implicit' in that tradition. Leavis crystallizes the Arnoldian inheritance of literature's 'centrality' and humanity, just as Richards concentrates the Arnoldian faith in poetry's consolatory 'concreteness' into a finished technique.

The apparent lapse into the very practicality scorned by Arnold is not the only curiosity in this period of English criticism, but an element in a series of strange reversals. The activity whose task was once seen as the defence of poetry has by the time of the Leavises become an offensive campaign (often in more senses than one) against neighbouring ideological territories. What had been seen by Coleridge as an endeavour necessarily requiring theoretical grounding has become a rallying-point for anti-theoretical sentiments. And the reading of poetry, which had according to Coleridge assumed a 'willing suspension of disbelief' in its fictions, was now valued for encouraging the suspension of belief in general. Within the topsy-turvy world of English criticism, this linked system of reversals merits the title of a literary-critical 'revolution', in a profoundly conservative and obscurantist direction. From the standpoint of that more 'general culture' which Arnold indicated as literary criticism's wider sphere of action, the most substantially damaging achievement of this revolution, and the greatest reversal of all, was its effect on the critical attitude in its widest sense. The critical approach, which refuses to accept what is offered simply at face value, which will not rest satisfied with things as they are, was squeezed into a narrowly literary criticism; social criticism in particular was blunted to conform with the implicit norms of literary 'sensibility' and put into the service of social consensus. The title of 'criticism' was usurped by a literary discourse whose entire attitude was at heart uncritical. Criticism in its most important and its most vital sense had been gutted and turned into its very opposite: an ideology.

NOTES

1. F. R. Leavis, '"Scrutiny": A Retrospect', *Scrutiny*, xx (Cambridge 1963), 4.

2. Ibid., 1. Cf. id., *Nor Shall My Sword* (1972), 63.

3. *Criterion*, ii (1927), 233. Arnold and Eliot would appear to have believed, with Shelley, that poetry 'is that which comprehends all science, and that to which all science must be referred'. Shelley, *Prose Works*, ii. 32.

4. *The Meaning of Meaning*, 159.

5. *Speculative Instruments*, 104.

6. *Complementarities*, 13.

7. See René Wellek, 'Literary Criticism and Philosophy', *Scrutiny*, v (1937), 376-83; F. R. Leavis, 'Literary Criticism and Philosophy: A Reply', *Scrutiny*, vi (1937), 59-70.

8. This methodological feature of post-war English criticism is examined fully in Pamela McCallum, 'The Cultural Theory of I. A. Richards, T. S. Eliot and F. R. Leavis 1922-1948: A Critique of Some Aspects of their Methodology and Assumptions' (Cambridge University Ph.D. thesis, 1978).

9. Coleridge, *Biographia Literaria*, ii. 116-17.

10. Ogden and Richards, *The Meaning of Meaning*, viii.

11. *After Strange Gods*, 13.

12. E. M. Forster, *Aspects of the Novel* (1927; rpt. Harmondsworth 1962), 21, 16.

13. *Complementarities*, 256; Brower *et al.* (eds.), *Richards*, 19.

14. F. R. Leavis, 'This Age in Literary Criticism', 8. The corollary of this rehabilitation was the purging from the canon of that radical tradition of English poets (Milton, the early Wordsworth, Shelley, Morris, Auden, and — except for the later Leavis — Blake) who celebrated the locomotive of history.

15. Watson, *Literary Critics*, 189.

16. F. R. Leavis, 'This Age', 8.

17. *Speculative Instruments*, 54; see also ibid., 108.

18. *Calendar of Modern Letters*, i. (1925), 70-1 (unsigned but since attributed to Edgell Rickword).

19. *Letters of Arnold to Clough*, ed. Lowry, 68 (1 March 1848).

20. '"Scrutiny", Retrospect', 6.

21. Ibid., 5-6.

22. In this instance, for example, Richards could intend either the original sense of the 'science of ideas' or the particular object of that science in the 'middle field' with which he is concerned in the previous paragraph.

23. F. R. Leavis (ed.), *Towards Standards of Criticism* (1933), 4.

24. 'This Age', 9.

25. Edmund Gosse, *Questions at Issue*, 180-2.

26. See, for example, John Gross, *The Rise and Fall of the Man of Letters* (1969), 281-2.

27. One of the essay questions set in *Culture and Environment* reads: 'The French Revolution established the peasant proprietor. If there had been an English Revolution at the same time, in what ways would England be different from what it is?' *CE*, 130.

28. Coleridge, *Biographia Literaria*, ii. 13.

29. *The Use of Poetry and the Use of Criticism*, 54.

30. '"Scrutiny", Retrospect', 2.

SELECT BIBLIOGRAPHY

Unless otherwise indicated, place of publication is London.

i. *Periodicals*

Calendar of Modern Letters, 4 vols., 1925-7.
Criterion, 18 vols., 1922-39; rpt. 1967.
Egoist: An Individualist Review, 6 vols., 1914-19.
Scrutiny: A Quarterly Review, 19 vols., Cambridge, 1932-53; 20th vol.
  ('Retrospect' and Indices), Cambridge, 1963.

ii. *Books and Pamphlets*

Annan, Noel, 'The Intellectual Aristocracy', in J. H. Plumb (ed.),
  Studies in Social History, 1955, 243-87.
Anon., *How to Teach English Literature*, 1880.
Anderson, Perry, 'Components of the National Culture', in A. Cockburn
  and R. Blackburn (eds.), *Student Power*, Harmondsworth, 1969,
  214-84.
Arnold, Matthew, *The Complete Prose Works of Matthew Arnold*, ed.
  R. H. Super, 11 vols., Ann Arbor, Michigan, 1960-77:
    i: *On the Classical Tradition*, 1960;
    ii: *Democratic Education*, 1962:
    iii: *Lectures and Essays in Criticism*, 1962;
    iv: *Schools and Universities on the Continent*, 1964;
    v: *Culture and Anarchy*, 1965;
    vi: *Dissent and Dogma*, 1968;
    vii: *God and The Bible*, 1970;
    viii: *Essays Religious and Mixed*, 1972;
    ix: *English Literature and Irish Politics*, 1973;
    x: *Philistinism in England and America*, 1974;
    xi: *The Last Word*, 1977.
——, *Culture and Anarchy*, ed. John Dover Wilson, Cambridge, 1932.
——, *Letters of Matthew Arnold 1848-1888* ed. George W. E. Russell,
  2 vols., 1895.
——, *The Letters of Matthew Arnold to Arthur Hugh Clough*, ed. Howard
  Foster Lowry, 1932.
——, *The Note Books of Matthew Arnold*, ed. Howard Foster Lowry,
  Karl Young, and Waldo Hilary Dunn, 1952.
——, *The Poems of Matthew Arnold*, ed. Kenneth Allott, 1965.
——, *Reports on Elementary Schools 1852-1882*, ed. Francis Sandford,
  1889.

——, *Unpublished Letters of Matthew Arnold*, ed. Arnold Whitridge, New Haven, 1923.

Baldwin, Stanley, *Presidential Address to the English Association*, 1928.

Bantock, G. H., 'Matthew Arnold, H.M.I.', *Scrutiny*, xviii, (1951), 32–44.

Barker, Ernest, *National Character and the Factors in its Formation*, 1927.

Bateson, F. W. *Essays in Critical Dissent*, 1972.

Beale, D. (ed.), *Reports Issued by the Schools Inquiry Commission on the Education of Girls*, 1869.

Belsey, Catherine, *Critical Practice*, 1979.

Benson, A. C. (ed.), *Cambridge Essays on Education*, Cambridge, 1917.

Bilan, R. P., *The Literary Criticism of F. R. Leavis*, Cambridge, 1979.

Board of Education, *Report of the Consultative Committee on The Education of the Adolescent* ('The Hadow Report'), 1926.

Boas, F. S. *Wordsworth's Patriotic Poems and their Significance Today*, 1914.

Boyers, Robert, *F. R. Leavis, Judgement and the Discipline of Thought*, Columbia, Missouri, 1978.

Brantlinger, Patrick, *The Spirit of Reform: British Literature and Politics 1832–1867*, 1977.

Brittain, F., *Arthur Quiller-Couch: A Biographical Study of Q*, Cambridge, 1948.

Brower, Reuben, Helen Vendler, and John Hollander (eds.), *I. A. Richards, Essays in His Honour*, 1973.

Brown, E. K., *Matthew Arnold: A Study in Conflict*, Chicago, 1948.

Bruford, W. H., *First Steps in German Fifty Years Ago*, 1965.

Buckley, J. H., *The Triumph of Time*, New Haven, 1966.

Buckley, Vincent, *Poetry and Morality: Studies in the Criticism of Matthew Arnold, T. S. Eliot and F. R. Leavis*, 1959.

Burnet, John, *Higher Education and the War*, 1917.

Burstyn, Joan N. *Victorian Education and the Ideal of Womanhood*, 1980.

Casey, John, *The Language of Criticism*, 1966.

Chace, W. M., *The Political Identities of Ezra Pound and T. S. Eliot*, Stanford, 1973.

Coleridge, S. T., *Biographia Literaria, or Biographical Sketches of My Literary Life and Opinions*, 2 vols., ed. J. Shawcross, 1907.

Collins, John Churton, *Ephemera Critica, or Plain Truths About Current Literature*, 1901.

——, *The Study of English Literature*, 1891.

Collins, L. C., *Life and Memoirs of John Churton Collins*, 1912.

Coulling, Sidney, *Matthew Arnold and His Critics: A Study of Arnold's Controversies*, Athens, Ohio, 1974.

Dale, P. A., *The Victorian Critic and the Idea of History*, Cambridge, Mass., 1977.

Davies, J. Llewelyn, *The Working Men's College 1854-1904*, 1904.

DeLaura, David J., *Hebrew and Hellene in Victorian England: Newman, Arnold, and Pater*, 1969.

de Selincourt, E., *English Poets and the National Ideal*, 1916.

——, *On Poetry*, Oxford, 1929.

——, *The Study of Poetry*, 1918.

Dingle, Herbert, *Science and Literary Criticism*, 1949.

Donovan, Robert A., 'The Method of Matthew Arnold's *Essays in Criticism*', *PMLA*, lxxi (1956), 922-31.

Dudley, F., 'Matthew Arnold and Science', *PMLA* lxxvii (1942), 275-94.

Eagleton, Terry, *Criticism and Ideology: A Study in Marxist Literary Theory*, 1976.

East India Company, Civil Service of, *Report to the Rt. Hon. Sir Charles Wood*, 1855.

Eliot, T. S., *After Strange Gods: A Primer of Modern Heresy*, 1934.

——, *For Lancelot Andrewes: Essays on Style and Order*, 1928.

——, *Homage to John Dryden: Three Essays on Poetry of the Seventeenth Century*, 1924.

——, Introduction to G. Wilson Knight, *The Wheel of Fire*, 1930.

——, 'A Note on Poetry and Belief', *Enemy*, i (1927), 15-17.

——, *Notes Towards the Definition of Culture*, 1948.

——, *On Poetry and Poets*, 1957.

——, *The Sacred Wood: Essays on Poetry and Criticism*, 1920, 7th edn. 1950.

——, *Selected Essays*, 1932, 3rd edn. 1951.

——, 'The Social Function of Poetry', *Adelphi* xxi (1945), 152-65.

——, *To Criticize the Critic*, 1965.

——, *The Use of Poetry and the Use of Criticism: Studies in the Relation of Criticism to Poetry in England*, 1933.

Faverty, F. E., *Matthew Arnold the Ethnologist*, Evanston, Illinois, 1951.

Fayolle, Roger, *Sainte-Beuve et le XVIIIe siècle, ou comment les Révolutions Arrivent*, Paris, 1972.

Fekete, Jôhn, *The Critical Twilight: Explorations in the Ideology of Anglo-American Literary Theory from Eliot to McLuhan*, 1977.

Filmer, Paul, 'The Literary Imagination and the Explanation of Socio-Cultural Change in Modern Britain', in *Archives Européennes de Sociologie*, x (1969), 271-91.

Firth, C. H., *The School of English Language and Literature: A Contribution to the History of Oxford Studies*, Oxford, 1909.

Fletcher, Ian, *Walter Pater*, Harlow, 1959, revd. 1971.

Freeman, E. A., 'Literature and Language', *Contemporary Review*, lii (1887), 549-67.

Gomme, Andor, *Attitudes to Criticism*, Carbondale, Illinois, 1966.

Gordon, G. S., *The Discipline of Letters, and Other Essays*, Oxford, 1946.

——, *The Letters of G. S. Gordon 1902-1942*, ed. M. C. Gordon, 1943.

——, *Poetry and the Moderns*, Oxford, 1935.

Gordon, M. C., *The Life of George S. Gordon 1881-1942*, 1945.

Gosse, Edmund, *Questions at Issue*, 1893.

Greenwood, Edward, *F. R. Leavis*, 1978.

Grierson, Herbert J. C. (ed.), *Metaphysical Poems and Lyrics of the Seventeenth Century*, Oxford, 1921.

Gross, John, *The Rise and Fall of the Man of Letters: English Literary Life Since 1800*, 1969; Harmondsworth, 1973.

Grylls, Rosalie Glynn, *Queen's College 1848-1948*, 1948.

Hales, J. W., 'The Teaching of English', in F. W. Farrar (ed.), *Essays on a Liberal Education*, 1867, 293-312.

Hamilton, Ian, *The Little Magazines: A Study of Six Editors*, 1976.

Harrison, Frederic, *The Choice of Books, and Other Literary Pieces*, 1903.

——, 'The Religious and Conservative Aspects of Positivism', *Contemporary Review*, xxvi (1875), 992-1012.

——, *Tennyson, Ruskin, Mill and Other Literary Estimates*, 1899.

Harrison, J. C. F., *A History of the Working Men's College 1854-1954*, 1954.

Harrison, John, *The Reactionaries*, 1967.

Hayman, Ronald, *Leavis*, 1976.

Health Exhibition Literature, The, *Vol XV: Conference on Education, Section C: Organisation of University Education*, 1884.

Herford, C. H., *The Bearing of English Studies upon the National Life*, 1910.

Hotopf, W. H. N., *Language, Thought and Comprehension: A Case Study in the Writings of I. A. Richards*, 1965.

Hough, Graham, *Image and Experience: Studies in a Literary Revolution*, 1960.

——, *The Last Romantics*, 1949.

Houghton, W. E., 'Victorian Anti-Intellectualism', *Journal of the History of Ideas* xiii (1952), 291-313.

Hurt, John, *Education in Evolution: Church, State, Society and Popular Education 1800-1870*, 1971.

Hyde, H. Montgomery, *The Trials of Oscar Wilde*, 1948, 2nd edn. Harmondsworth, 1962; New York, 1973.

Hyman, Stanley Edgar, *The Armed Vision: A Study in the Methods of Modern Literary Criticism*, New York, 1948.

James, D. G., *Matthew Arnold and the Decline of English Romanticism*, Oxford, 1961.

Kampf, Louis, and Paul Lauter (eds.), *The Politics of Literature*, New York, 1973.

Kenner, Hugh, *The Invisible Poet: T. S. Eliot*, 1960.

Kermode, Frank, *Romantic Image*, 1957.

Kingsley, Charles, *Literary and General Lectures and Essays*, 1880.

——, *The Roman and the Teuton*, 1877.

Knights, Ben, *The Idea of the Clerisy in the Nineteenth Century*, Cambridge, 1978.

Knox, Vicesimus (ed.), *Elegant Extracts*, 1824.

Kojecký, Roger, *T. S. Eliot's Social Criticism*, 1971.

Kolakowski, Leszek, *Positivist Philosophy from Hume to the Vienna Circle*, Harmondsworth, 1972.

Kramer, Judith R., 'The Social Role of the Literary Critic', in M. C. Albrecht (ed.), *The Sociology of Literature and Art*, 1970, 437-54.

Kreiger, Murray, 'The Critical Legacy of Matthew Arnold, or the Strange Brotherhood of T. S. Eliot, I. A. Richards and Northrop Frye', *Southern Review*, v (1969), 457-75.

Lawford, Paul, 'Conservative Empiricism in Literary Theory: A Scrutiny of the Work of F. R. Leavis', *Red Letters*, no. 1, n. d., 12-15; no. 2, Summer 1976, 9-11.

Leavis, F. R., *The Common Pursuit*, 1952.

——, 'Criticism of the Year', *Bookman*, lxxxi (1931), 180.

——, (ed.), *Determinations: Critical Essays*, 1934.

——, *Education and the University: A Sketch for an 'English School'*, 1943.

——, *English Literature in Our Time and the University*, 1969.

——, *For Continuity*, Cambridge, 1933.

——, *How to Teach Reading: A Primer for Ezra Pound*, Cambridge, 1932.

——, *Letters in Criticism*, ed. John Tasker, 1974.

——, *New Bearings in English Poetry: A Study of the Contemporary Situation*, 1932, 2nd edn. 1950.

——, *Nor Shall My Sword: Discourses on Pluralism, Compassion and Social Hope*, 1972.

——, *Revaluation: Tradition and Development in English Poetry*, 1936.

——, (ed.), *A Selection from Scrutiny*, 2 vols., Cambridge, 1968.

——, 'This Age in Literary Criticism', *Bookman*, lxxxiii (1932), 8-9.

——, (ed.), *Towards Standards of Criticism*, 1933.

——, and Denys Thompson, *Culture and Environment: The Training of Critical Awareness*, 1933.

Leavis, Q. D. *Fiction and the Reading Public*, 1932, Harmondsworth, 1979.

McCallum, Pamela, 'The Cultural Theory of I. A. Richards, T. S. Eliot and F. R. Leavis 1922-1948: A Critique of Some Aspects of their Methodology and Assumptions', unpublished Ph.D. thesis, Cambridge, 1978.

McCarthy, Patrick J., *Matthew Arnold and the Three Classes*, New York, 1964.

Macaulay, Thomas Babington, *Speeches on Politics and Literature by Lord Macaulay*, 1909.

Macherey, Pierre, *A Theory of Literary Production*, 1978.

McKenzie, D. F., and M. -P. Allum, *F. R. Leavis: A Check-List 1924-1964*, 1966.

Madden, W. A., 'The Divided Tradition of English Criticism', *PMLA*, lxxiii (1958), 69-80.

Malone, David H. (ed.), *The Frontiers of Literary Criticism*, Los Angeles, 1974.

Martin, Graham (ed.), *Eliot in Perspective*, 1970.

Mathieson, Margaret, *The Preachers of Culture: A Study of English and its Teachers*, 1975.

Maurice, F. D., *The Friendship of Books and Other Lectures*, 1874.

——, *Lectures to Ladies on Practical Subjects*, 1855.

Milner, Andrew, 'Leavis and English Literary Criticism', *Praxis* (Berkeley), i (1976), 91-106.

Mulhern, Francis, *The Moment of 'Scrutiny'*, 1979.

Needham, John, *The Completest Mode: I. A. Richards and the Continuity of English Literary Criticism*, Edinburgh, 1982.

Newbolt, Henry, *The Idea of an English Association*, 1928.

——, *The World as in My Time*, 1932.

Newbolt, Margaret (ed.), *The Later Life and Letters of Sir Henry Newbolt*, 1942.

Newman, Francis W., Translator's Preface to V. Huber, *The English Universities*, 2 vols. 1843.

Newton-de-Molina, D. (ed.), *The Literary Criticism of T. S. Eliot: New Essays*, 1977.

Oakley, John, 'The Boundaries of Hegemony: Pater', in Francis Barker *et al.* (eds.), *Literature, Society and the Sociology of Literature*, University of Essex, 1977.

Ogden, C. K., and I. A. Richards, *The Meaning of Meaning*, 1923.

——, I. A. Richards, and James Wood, *The Foundations of Aesthetics*, 1922.

Palmer, D. J. *The Rise of English Studies*, 1965.

Parrinder, Patrick, *Authors and Authority: A Study of English Literary Criticism and its Relation to Culture 1750-1900*, 1977.

——, 'Sermons, Pseudo-Science and Critical Discourse: some reflections on the aims and methods of contemporary English', *Studies in Higher Education*, iv (1979), 3-13.

Pater, Walter, *Appreciations, with an Essay on Style*, 1889; Library Edition 1910.

——, *Imaginary Portraits*, 1887; Library Edition 1910.

——, *Marius the Epicurean: His Sensations and Ideas*, 2 vols., 1885; Library Edition 1910.

——, *Plato and Platonism*, 1893; Library Edition 1910.

——, *Selected Writings of Walter Pater*, ed. Harold Bloom, New York 1974.

——, *Studies in the History of the Renaissance*, 1873; 2nd and subsequent edns. retitled *The Renaissance: Studies in Art and Poetry*, Library Edition 1910.

Peckham, M., *Victorian Revolutionaries: Speculations on Some Heroes of a Culture Crisis*, New York, 1970.

Potter, Stephen, *The Muse in Chains: A Study in Education*, 1937.

Quiller-Couch, Arthur, *Cambridge Lectures*, 1943.

——, *On the Art of Writing*, Cambridge, 1916.

——, *Studies in Literature*, Cambridge, 1918.

——, *Studies in Literature: Third Series*, Cambridge, 1929.

Raleigh, Walter, *The Letters of Sir Walter Raleigh 1879-1922*, ed. Lady Raleigh, 2 vols., 1926.

——, *On Writing and Writers*, ed. G. S. Gordon, 1926.

——, *Some Authors*, Oxford, 1923.

——, *Style*, 1897.

Richards, I. A., *Complementarities: Uncollected Essays*, ed. J. P. Russo, Manchester, 1977.

——, *Interpretation in Teaching*, 1938.

——, *The Philosophy of Rhetoric*, New York, 1936.

——, *Practical Criticism: A Study of Literary Judgement*, 1929.

——, *Principles of Literary Criticism*, 1924; 3rd edn. 1928; reset 1967.

——, *Science and Poetry*, 1926.

——, *Speculative Instruments*, 1955.

Rickword, Edgell, *Essays and Opinions 1921-1931*, ed. Alan Young, Cheadle, Cheshire, 1974.

Robbins, R. H. *The T. S. Eliot Myth*, New York 1951.

Roberts, R. D. (ed.), *Aspects of Modern Study*, 1894.

Robertson, P. J. M., *The Leavises on Fiction*, 1981.

Robinson, H. G. 'On the Use of English Classical Literature in the Work of Education', *Macmillan's Magazine*, ii (1860), 425-34.

Rolph, C. H. (ed.), *The Trial of Lady Chatterley: Regina v. Penguin Books Limited*, Harmondsworth, 1961.

Sainte-Beuve, C. -A., *Causeries du lundi*, 15 vols., 3rd edn., Paris, 1857-72.

Sampson, G., *English for the English*, 1921; 2nd edn. 1925.

Schiff, Hilda (ed.), *Contemporary Approaches to English Studies*, 1977.

Schiller, Jerome P., *I. A. Richards' Theory of Literature*, New Haven, 1969.

Schuchard, Ronald, 'Eliot and Hulme in 1916: Towards a Revaluation of Eliot's Critical and Spiritual Development', *PMLA*, lxxxviii (1973), 1083-94.

——, 'T. S. Eliot as Extension Lecturer 1916-1919', *Review of English Studies*, xxv (1974), 163-73, 292-304.

Seton-Watson, R. W., J. Dover Wilson, Alfred E. Zimmern, and Arthur Greenwood, *The War and Democracy*, 1914.

Shelley, Percy Bysshe, *Prose Works*, 2 vols., ed. R. H. Shepherd, 1912.

Sidgwick, Henry, *Miscellaneous Essays and Addresses*, 1904.

——, 'The Prophet of Culture', *Macmillan's Magazine*, xvi (1867), 274-80.

Simon, W. M., 'Auguste Comte's English Disciples', *Victorian Studies*, viii (1964), 161-72.

Smith, D. Nichol, *The Functions of Criticism*, Oxford, 1909.

Smith, Nowell, *The Origin and History of the Association*, 1942.

Spurgeon, C. F. E., *Poetry in the Light of War*, 1917.

Stead, C. K., *The New Poetic: Yeats to Eliot*, 1964.

Steiner, George, *Language and Silence*, 1967.

Super, R. H., 'Documents of the Matthew Arnold–Sainte-Beuve Relationship', *Modern Philology*, lx (1963), 206-10.

Swingewood, Alan, *The Myth of Mass Culture*, 1977.

The Teaching of English in England, 1921.

Tillyard, E. M. W., *Essays Literary and Educational*, 1962.

——, *The Muse Unchained: An Intimate Account of the Revolution in English Studies at Cambridge*, 1958.

Trilling, Lionel, *Beyond Culture: Essays on Literature and Learning*, 1966.

——, *A Gathering of Fugitives*, 1957.

——; *Matthew Arnold*, New York, 1939; 3rd edn. 1949.

Trotter, W., *Instincts of the Herd in Peace and War*, 1916; 2nd edn. 1919.

Verma, Rajandra, *Royalist in Politics: T. S. Eliot and Political Philosophy*, 1968.

Vogeler, M. S., 'Matthew Arnold and Frederic Harrison: The Prophet of Culture and the Prophet of Positivism', *Studies in English Literature*, ii (1962), 441-62.

Walcott, F. G., *The Origins of 'Culture and Anarchy'*, 1970.

Walsh, William, *F. R. Leavis*, 1980.

——, *The Use of Imagination: Educational Thought and the Literary Mind*, 1960.

Ward, Anthony, *Walter Pater: The Idea in Nature*, 1966.

Warren, Alba, *English Poetic Theory 1825-1865*, Princeton, 1950.

Watson, Garry, *The Leavises, the 'Social' and The Left*, Swansea, 1977.

Watson, George, *The English Ideology*, 1973.

——, *The Literary Critics: A Study of English Descriptive Criticism*, 1962; 2nd edn. Harmondsworth, 1973.

——, 'The Triumph of T. S. Eliot', *Critical Quarterly*, vii (1965), 328-37.

Weimann, Robert, *Structure and Society in Literary History*, 1977.

Wellek, René, *A History of Modern Criticism 1750-1950:* vol. 3: *The Age of Transition*; vol. 4: *The Later Nineteenth Century*, New Haven, 1965.

West, Alick, *Crisis and Criticism*, 1937.

Whitridge, Arnold, 'Matthew Arnold and Sainte-Beuve', *PMLA*, liii (1938), 303-13.

Widdowson, Peter (ed.), *Re-Reading English*, 1982.

Willey, Basil, *Cambridge and Other Memories*, 1964.

Williams, Raymond, *Culture*, 1981.

——, *Culture and Society 1780-1950*, 1958.

——, *Keywords: A Vocabulary of Culture and Society*, 1976.

Wilson, John Dover, *Milestones on the Dover Road*, 1969.

——, (ed.), *The Schools of England: A Study in Renaissance*, 1928.

——, *What Happens in Hamlet*, Cambridge, 1935.

Wimsatt, W. K. Jr., and Cleanth Brooks, *Literary Criticism: A Short History*, New York, 1957.

Woolf, Virginia, *A Room of One's Own*, 1929.

——, 'Walter Raleigh', in id., *Collected Essays*, i (1966), 314-18.

Wright, D. G., 'The Great War, Government Propaganda, and English "Men of Letters" 1914-1916', *Literature and History*, no. 7 (Spring 1978), 70-100.

Wright, Iain, 'F. R. Leavis, the "Scrutiny" Movement and the Crisis', in Jon Clark, Margot Heinemann, David Margolies, and Carole Snee (eds.), *Culture and Crisis in Britain in the 'Thirties*, 1979, 39-65.

Addendum, 1987

Since the above bibliography was compiled in 1982, Pamela McCallum's thesis has been published as *Literature and Method: Towards a Critique of I. A. Richards, T. S. Eliot and F. R. Leavis* (1983). The following works of related interest have also appeared:

Eagleton, Terry, *The Function of Criticism: From the Spectator to Post-structuralism*, 1984.

——, *Literary Theory: An Introduction*, Oxford, 1983.

Hawkes, Terence, *That Shakespeherian Rag: Essays on a Critical Process*, 1986.

McMurtry, Jo, *English Language, English Literature: The Creation of an Academic Discipline*, Hamden, Connecticut, 1985.

Williams, Raymond, *Writing in Society*, 1983.

INDEX

Académie Française, 44-6
advertising, 140, 166-7, 185-92, 194, 222
Aeschylus, 20, 101
Anglicanism, 39, 46, 109, 122, 229
anthropology, 80, 176, 182, 201-2
anti-intellectualism, 24, 42
Aristophanes, 9
Aristotle, 81, 134, 156, 165, 201
Arnold, Matthew, 3, 4, 6, 9-11, 14-15, 18-57, 59-60, 63, 67, 72-4, 78, 80-1, 83, 87-8, 92, 94-5, 97-102, 105, 109, 111-21, 123-5, 127-32, 137, 139, 141, 143, 145-8, 157, 160, 162-4, 168, 174, 176, 179, 192-4, 196, 198-201, 203-4, 206-14, 217-34; works: 'Bishop and the Philosopher', 31, 42; *Culture and Anarchy*, 35-6, 44, 45, 47, 49, 51-3, 55, 83, 95, 99-101, 107, 176, 207, 218; 'Equality', 36, 101; *Essays in Criticism*, 31-2, 46, 97, 113, 210; 'Function of Criticism . . .', 19-25, 32, 41, 119; 'Literary Influence of Academies', 44-5, 143; *Literature and Dogma*, 37, 213; 'Literature and Science', 40-1, 164; 'Modern Element in Literature', 28-9, 123, 211, 219; *On the Study of Celtic Literature*, 36, 132, 219; *On Translating Homer*, 29-32; Preface to *Poems*, 26-8, 115; 'Scholar Gipsy', 48, 198; 'Study of Poetry', 18-20, 39-40, 223, 225
Arnold, Thomas, 28, 33, 60
Auden, W. H., 217, 235

Baldwin, Stanley, 103
Beaverbrook, Lord, 87
Bennett, Arnold, 87, 164, 166
Blake, William, 78, 128--9, 132, 235
Boas, F. S., 88, 94
Bonaparte, Louis, 13
Bonaparte, Napoleon, 1
Bourne, George, 173, 192-3
Bradley, A. C., 79, 87, 198

Bradley, F. H., 124
Brecht, Bertolt, 8
British Honduras, 185-6
Brontë, Charlotte, 178, 182
Brooke, Rupert, 86
Browning, Robert, 122-3
Brydges, Egerton, 179
Bunyan, John, 173, 176-8, 192, 220
Burke, Edmund, 210
Burnet, John, 88
Burroughs, Edgar Rice, 164
Byron, Lord, 5, 26, 39, 92

Calendar of Modern Letters, 220-1, 228
Cambridge, 61, 69, 71, 80, 88, 106, 134-5, 149, 155-6, 158-9, 162, 175, 196-8, 211
Carlyle, Thomas, 25, 26
Catholicism, 32, 51-2, 110, 129, 231
Chadwick, H. M., 80, 159
Chartism, 63, 66, 83
Chaucer, Geoffrey, 39
Church of England, *see* Anglicanism
cinema, 118, 131, 138-9, 159, 163, 166, 170, 181-2, 185, 217, 222, 225
Civil Service of East India Co., *see* India Civil Service
Civil War (English Revolution), 10, 121-2, 125, 231
Clough, A. H., 15, 26
Cobbett, William, 25, 192
Colenso, William, 31, 32, 37, 42, 206
Coleridge, S. T., 9, 10, 45, 88, 115, 126, 132, 141, 160, 208, 213, 232-4
Collins, J. C., 61, 64, 71, 72-5, 200
communism, 98, 99, 102-3, 171, 175
Comte, Auguste, 51
Cooper, Thomas, 66
Criterion, 131, 168, 190

Dante Alighieri, 59, 127, 128, 132, 164, 227
Darwin, Charles, 41, 165, 216, 218-19
Defoe, Daniel, 177, 178

de Selincourt, Ernest, 91-2, 98, 107
Dickens, Charles, 178, 179, 182
dissociation of sensibility, 3, 121-2, 125, 132, 176, 214, 217
Donne, John, 113, 122-3, 164
Dos Passos, John, 171, 173, 188
Dryden, John, 6, 7, 9, 10, 12, 26, 113, 122, 162

Egoist, 200
Eliot, George, 51
Eliot, T. S., 3, 4, 6, 9, 15, 24-7, 49, 51, 53, 56, 86, 109-34, 137, 141, 156-8, 162, 164, 168, 172, 176-7, 196-8, 200-2, 204, 206, 208-14, 217, 219, 222, 226-7, 231, 233; works: After Strange Gods, 118, 120, 212; 'Andrew Marvell', 124-6; For Lancelot Andrewes, 109, 122; 'Function of Criticism', 113, 116, 118; Sacred Wood, 25, 109, 112-18, 125-30, 158, 200-2; 'Tradition and the Individual Talent', 115, 118, 120-1, 200, 212; Use of Poetry, 6-7, 24, 156-8, 162; Waste Land, 111, 123, 136, 156-7
Empson, William, 135, 197, 204
English Association, 93-4, 103-4, 168, 196, 221
extension lectures, 61, 63, 64, 67-8, 72, 74-5, 95, 130, 133

Fielding, Henry, 5, 163
Firth, C. H., 94
Forster, E. M., 211
Frazer, J. G., 112
Freeman, E. A., 73-5, 155
French Revolution, 21, 30-1, 35, 211, 236
Freud, Sigmund, 165, 216-17

Galsworthy, John, 87
General Strike, 99, 100, 103
Goethe, J. W. von, 26, 59, 126, 132
Gordon, G. S., 89, 104-6, 111, 196
Gosse, Edmund, 179, 210, 229-30
Grierson, H., 132

Hadow Report, 103
Hales, J. W., 62
Hardy, Thomas, 87, 163, 164
Harmsworth, H., 80, 87

Harrison, Frederic, 51
Hebraism, 36, 49-50, 218
Hegel, G. W. F., 28, 56
Hellenism, 36, 49-50, 51, 72, 73, 80, 124, 218
Homer, 29-30, 59, 72, 101
Hughes, Thomas, 63, 65, 83
Hulme, T. E., 110
Huxley, Thomas, 37, 51

India Civil Service, 61, 70-2

James, Henry, 112, 183
Johnson, Samuel, 7, 11, 121, 162, 227
Jonson, Ben, 233
Joyce, James, 86, 111, 164, 183, 188

Keats, John, 5, 38, 76, 81
Ker, W. P., 94
Kingsley, Charles, 68-9, 74, 84, 214
Knight, G. Wilson, 198
Knights, L. C., 198
Knox, Vicesimus, 59

'Lady Chatterley' trial, 8
Lawrence, D. H., 86, 172-4, 183, 188, 205, 212, 215, 217, 231
Leavis, F. R., 3, 4, 16, 24-5, 42, 56, 86, 96, 132, 162-75, 177, 179-81, 183-4, 186-94, 196-8, 203-5, 207, 209-13, 215-17, 220, 222, 225-8, 230-1, 233-4; works: Culture and Environment, 162, 175, 180, 186-93, 207, 212, 220, 225, 236; Great Tradition, 162; For Continuity, 162, 167-75, 212; Mass Civilisation, 162, 165-8, 177, 179, 181, 188; New Bearings, 163
Leavis, Q. D., 4, 16, 79, 96, 162, 175-86, 188-9, 194, 197, 203-6, 209-12, 215-17, 220, 222, 225-7, 230-1; Fiction and the Reading Public, 162, 175-86, 206-7
Lloyd George, David, 93
Lukács, Georg, 9

Macaulay, T. B., 63, 70-1, 76
Malinowski, B., 194
Marvell, Andrew, 125
Marx, Karl, 99, 165, 171, 191
Marxism, 1, 2, 169-74, 188-9, 191, 198, 209

Masson, David, 84
Masterman, C. F. G., 87
masturbation, 182, 188, 206, 217
Maurice, F. D., 63-4, 68-9, 74, 84
Mill, John Stuart, 51
Milton, John, 66, 81, 91-2, 101, 122, 129, 220, 235; *Paradise Lost*, 177, 224
Morley, John, 67
Morris, William, 79, 92, 235
Mulhern, Francis, 17, 194
Murry, J. M., 117

Newbolt, Henry, 65, 87, 93, 95, 103-4, 111, 219, 221
Newbolt Report, 89-90, 92-107, 137, 147, 159, 160, 196, 206, 225, 229-31
Newcastle Commission, 62
Newman, Francis, 30-1
Newman, John Henry, 48
Northcliffe, Lord, 179

Ogden, C. K., 135, 202, 208
Owen, Wilfred, 86
Oxford, 46-9, 59-61, 67, 69-74, 75-7, 80, 88, 104, 106, 109, 155, 196-8, 207, 210

Palmer, D. J., 83, 84
Parrinder, Patrick, 6
Pater, Walter, 4, 49-57, 59, 133, 203, 213-14, 231
philology, 73-4, 80, 89, 106, 200
Plato, 9, 10, 218
Positivism, 51-3
Pound, Ezra, 106, 110-11, 113, 115, 117, 164-5, 167-8, 173, 215, 226
practical criticism, 4, 23, 57, 155-6, 159-60, 187, 197-8, 211, 215, 232-4
Protestantism, 32, 48, 83, 125, 155, 210-11, 231
psychoanalysis, 115, 153, 215-17

Quiller-Couch, Arthur, 69, 80-2, 87-9, 94, 107, 134, 160, 200, 204, 219
Queen's College for Women, 68-9, 84

rabbits, 112, 131
race, 84, 178, 218-21

Raleigh, Walter 72, 76-80, 82, 85, 88-9, 92, 104-6, 196, 200, 208, 214
Revolution, English, *see* Civil War
Revolutions of 1848, 13-15, 17
Richards, I. A., 4, 16, 24, 56, 75, 94, 134-66, 168, 172, 175-6, 183, 187-8, 192, 196-8, 202-6, 208-20, 222, 224-6, 228-31, 233; works: *Meaning of Meaning*, 135-6, 194, 202-3, 208; *Practical Criticism*, 135, 137, 141, 147, 149-55, 158, 204, 219, 226, 228, 233; *Principles of Literary Criticism*, 56, 135, 137-8, 148-9, 152, 175, 224; *Science and Poetry*, 135, 137, 145, 164, 202, 216
Richardson, Samuel, 178
Robinson, H. G., 60, 62, 65-7, 229
Romanticism, 11, 14, 18, 21, 26, 27, 78, 110, 112, 113, 117, 124, 128, 156, 201
Rothermere, *see* Harmsworth
Rousseau, J.-J., 110
Ruskin, John, 25
Russell, Bertrand, 171

Sainte-Beuve, C.-A., 10-15, 22, 26, 160-1, 218
Saintsbury, George, 6, 79, 134-5
Saint-Simon, C. H., 97
Sampson, George, 98, 100-3, 105, 108, 147, 159, 221
science, 19, 40-1, 62, 73, 78, 146, 164, 171, 199-203, 207
Scrutiny, 16, 86, 162, 169, 173-4, 187, 196, 198, 204, 209, 220, 222, 225, 230, 233-4
Shakespeare, William, 5, 20, 28, 66, 77, 78, 79, 81, 82, 84, 92, 101, 105, 107, 163, 164, 174, 177, 198, 212, 220, 229; works; *Hamlet*, 115-16; *King Lear*, 143; *Measure for Measure*, 50
Shaw, George Bernard, 127, 171
Shelley, Percy Bysshe, 38, 73, 75, 92, 101, 142, 160, 166, 224, 235
Sidgwick, Henry, 62
Smith, Nowell, 83, 93
Sophocles, 28, 29, 59, 101
Spinoza, B. de, 31, 122, 132
Spurgeon, Caroline F. E., 90-1, 94, 98, 214

structuralism, 1-3
Symons, Arthur, 133, 134-5

Taunton Commission, 63, 68-9, 72,
74
Tennyson, Alfred, 122-3
Teutonism, 84, 87-9, 219
Thompson, Denys, 156, 180, 186,
187, 194
Tillyard, E. M. W., 61, 155, 156,
159, 197
Tocqueville, Alexis de, 210
Trilling, Lionel, 35, 193
Trotsky, Leon, 175
Trotter, William, 180-1, 182, 194,
220

Vico, Giambattista, 28

Watson, George, 6, 9, 112, 131-2, 215

Wellek, René, 205
Wells, H. G., 87, 164, 171
Wilde, Oscar, 8, 50, 54, 56
Willey, Basil, 86, 87, 89, 134, 135,
197, 202
Williams, Raymond, 132
Wilson, John Dover, 94, 98-100,
107-8
women's education, 61, 67-9, 72,
74
Woolf, Virginia, 79, 85, 164
Wordsworth, William, 5, 7, 10, 39,
78, 88, 126, 142, 174, 235
Working Men's College, 61, 63-4
World War I, 4, 60, 79, 80, 86-95,
104, 106, 111-12, 134-6, 141,
159, 186, 210

Yeats, W. B., 53, 111